Work Mate Marry Love

Work

Mate

Marry

Love

How Machines Shape Our Human Destiny

DEBORA L. SPAR

FARRAR, STRAUS AND GIROUX • NEW YORK

Farrar, Straus and Giroux
120 Broadway, New York 10271

Grateful acknowledgment is made for permission to reproduce the following
material:
"The River" by Bruce Springsteen. Copyright © 1980 Bruce Springsteen
(Global Music Rights). Reprinted by permission. International copyright
secured. All rights reserved.
"If I Should Fall Behind" by Bruce Springsteen. Copyright © 1992 Bruce
Springsteen (Global Music Rights). Reprinted by permission. International
copyright secured. All rights reserved.

Owing to limitations of space, illustration credits can be found on page 369.

Library of Congress Cataloging-in-Publication Data
Names: Spar, Debora L., author.
Title: Work mate marry love : how machines shape our human destiny /
    Debora L. Spar.
Description: First edition. | New York : Farrar, Straus and Giroux, 2020. |
    Includes bibliographical references and index.
Identifiers: LCCN 2020012314 | ISBN 9780374200039 (hardcover)
Subjects: LCSH: Technological innovations—Social aspects. | Families—
    History. | Man-woman relationships—History. | Work—History. |
    Social change. | Technology and civilization.
Classification: LCC HM846 .S693 2020 | DDC 303.4—dc23
LC record available at https://lccn.loc.gov/2020012314

Designed by Richard Oriolo

Our books may be purchased in bulk for promotional, educational, or
business use. Please contact your local bookseller or the Macmillan Corporate
and Premium Sales Department at 1-800-221-7945, extension 5442, or by e-mail
at MacmillanSpecialMarkets@macmillan.com.

www.fsgbooks.com
www.twitter.com/fsgbooks • www.facebook.com/fsgbooks

10   9   8   7   6   5   4   3   2   1

To the memory of Mary Margaret Winifred Billington,
whose indomitable spirit endures

First we build the tools, then they build us.
—Marshall McLuhan

Contents

Work Mate Marry Love

Prologue
The Futures of the Past

The future of the past was a bright and shiny thing.

In 1851, when Britain's Prince Albert triumphantly opened his country's Great Exhibition in the Crystal Palace, he saw it as representing an inflection point in history, a moment when humankind would come together to praise the marvels of technology—engines and looms and printing machinery—and contemplate the wonders ahead. The exhibition took place in a castle made of glass, three hundred thousand sheets of the largest expanses of glass ever made, plus spans of cast-iron columns and a massive curved roof hovering above a canopy of live trees. It was an "Arabian Night's structure . . . a splendid phantasm," a glimpse of a future that had just been born.[1]

Fast-forward now to the World's Fair of 1964, hosted by an America bursting with progress and eager to revel in the joys of tomorrow. Like its Victorian predecessor, the 1964 fair was organized by local luminaries and dedicated, ostensibly, to peace, understanding, and the advent of technology. But it was technology that really stole the show again, captivating visitors with its seductive vision of what the future would bring. There were films from IBM on the inner workings of computers, and an array of full-scale rocket models. There was a Picturephone, conveying callers' images along with their voices. There were lunar colonies and microwave ovens and cars that could travel through space. In the Festival of Gas pavilion, a Norge Dish Maker washed the family of the future's plastic dinnerware, and then ground it into pellets that could be transformed into new plates and saucers.[2] "It was almost," said one observer, "like a dream world."[3]

The funny thing about the future, though, is how new it actually is. Until recently, after all, humans couldn't really imagine a future that was any different from the past. For thousands and thousands of years, across every corner of the world, people lived more or less as their parents and grandparents had. Individuals faced the tiny details of their fate—the person they'd mate with and the children they bore; the length of their lives and the way they died—but their lives in a broader sense would unfurl across the exact same landscape as their ancestors'. The same modes of transportation. The same kinds of homes. The same ways of sowing and reaping and praying for rain. When nothing changed from year to year, the very idea of the future was inconceivable.

It was technology that changed this world, and created a future that was different from the past. Because once technology crashed into people's lives—once it upended the age-old ways of farming or building or moving—people living through any particular moment in time could imagine later moments that were different, and better. People could imagine change and progress, a world ahead that was different—faster, sleeker, sexier, richer—than what they had always known. They could begin to dream of that future, and conceive the tools that would roust it into being.

In dreaming of this shiny future, though, we humans have demonstrated an uncanny knack for compartmentalization, neatly dividing our workaday selves from our more intimate identities. When we think about technology, we think about machines. Steam engines and spinning looms in Prince Albert's day, robots and drones and driverless cars in ours. And when we think about our future selves, we presume that our tools will change, but not the inner currents of our lives. In other words, even when the pace of technological change is racing and accelerating, even when we can easily glimpse a future marked by artificial intelligence and casual rides to Mars, we still somehow imagine that we—as a species and as individuals—will remain largely the same. That we'll have sex and fall in love; marry, bear children, and die. That we'll live amid the bright and shiny tools of technology, but not be altered by them in any fundamental way.

It is tempting to feel this way, and to imagine a future that feels much like the past—same families, same yearnings, same mixed-up relationships and enduring loves. But this just can't be right. Because as technology changes, it is going to change us as well. Indeed, it already has.

In particular, the technologies of the twenty-first century are already beginning to hammer away at what have long been the contours of our romantic and family lives. How we fall in love. How we have sex and bear children. How we think, at the most basic level, of our bodies and ourselves. The old normal—the normal that endured for so long that we presumed it to be eternal—is already starting to morph and evolve, paving the way toward a fundamentally reconfigured future.

Let's start with the most basic foundation of family life: babies. It used to be that they were born through standard, predictable channels—

created by one man and one woman, usually living together in a legally sanctioned union known as marriage. This was the structure that surrounded and legitimized the inherently human act of reproduction, the structure that was celebrated, blessed, and repeated endlessly across nearly all the world's developed societies. Although marriage has never been a *biological* prerequisite for babies, it has—for ages—been the context that surrounds them, a context so prevalent and pervasive as to feel carved in stone.

But it's not. In fact, in the course of less than fifty years, the norms of marriage and baby-making have already undergone a seismic set of changes. In 2015, nearly half the children in the United States were living in nontraditional households, and 40 percent of births were to unmarried women—up from only 5 percent in 1960. In South America, likewise, more than half of all children are currently born to unmarried mothers, as are between one-third and one-half of children in western Europe. Across the Western world, as a result, the nuclear family is the norm in a declining number of households, replaced by a growing hodgepodge of single-parent households, cohabiting adults, and extended stepparenting arrangements.

Outside the family structure, similarly radical changes are transforming sex and gender and love. Same-sex marriage—virtually unimaginable as recently as fifty years ago—has surged from the shadows to become both acceptable and celebrated, and gay weddings now regularly grace the pages of most metropolitan newspapers. Gay couples are having and adopting children in numbers that would have been literally inconceivable just a few decades earlier, and children, regardless of their parentage, are increasingly being hatched from a growing array of high-technology means: donor sperm, donor eggs, frozen eggs and embryos. Some of these children, moreover—some of the generation born around the turn of the new millennium—are also deciding as they grow up to transition from one gender to another, transforming their own bodies in the process, along with society's long-standing notions of sex and gender and identity.

As these changes have rocketed through the world's social infrastructure so, too, have technological innovations been changing the face of industry and commerce. Smartphones, 3D printing, the entire Uber-centric economy of a technology-enabled transient labor force—they all burst into commercial prominence between 2000 and 2015, heralding a new machine age and delivering vast wealth to those along its cutting edge. No one knows, really, how these technologies will evolve over time, and how they will shape the workplace and workforce of the future. But we know the changes will be vast, and their impact—like that of most technological revolutions—profound.

What is less frequently commented upon, though, are the deep connections between the machines that are emerging from this wave of innovation and the social changes occurring as a result. In other words, we tend, both casually and in more formal examinations, to put technology (the cars, the phones, the robots) into one bucket and social change (gay marriage, gender transitions, the disappearing nuclear family) into another. We presume that we'll get those flying cars and robotic maids but will otherwise live in the future pretty much exactly as we do today.

It's an enticing thought—imagining our future selves as better-equipped models of our present—but it's wrong. Instead, today's technological changes will inevitably and inherently transform not just the worlds of business and commerce and industry, but the realms of love, sex, and family as well. We will change, as we always have, along with our machines.

WORK MATE MARRY LOVE explores the intimate and fundamental nature of these transformations, tracing how our social structures—how we work and love, the ways in which we build our families and give birth to our children—are shaped and created by the technologies that prevail during a particular era. Monogamous marriage, for example, was not the norm in our distant past; instead, it emerged along with the development of agricultural tools. The nuclear family is not a biological given, but rather a distinct creation of the industrial age, facilitated by a range of technological innovations that drew men into newly created factories and women to newly complicated homes. In those parts of the world where the Industrial Revolution was late or absent (such as Guinea and Burkina Faso, for example), polygamy is still more commonly practiced.[4] In those parts where technology is accelerating, pushing humankind to a postindustrial age, family forms will morph again. They are already doing so.

To trace these developments, this book ranges across several broad swaths of time, focusing on crucial moments of innovation and the social changes they ignited. It starts in the ancient past, with the agricultural technologies—most notably the plow—that heralded civilization's rise. Today the plow seems almost laughably old-fashioned, a rickety contraption of wooden poles and blades, yoked to an ox. When the plow emerged in the delta of the Tigris and Euphrates Rivers, though, it was revolutionary, allowing once-nomadic people to turn from foraging to farming, and to develop what would soon become villages, and then towns and cities. Civilization emerged from the population density that sedentary farming afforded, as did our modern notions of both the state and private property. Before the plow, land use was communal, and people moved as tribes. After its invention, individuals began to claim specific plots and implements as their own. My farm. My cow. My stores of seed and wheat and flour. Crucially, some of these individuals—the men—began to claim others—the women—as their property as well. It wasn't pride that made them do this, or physical prowess, but rather the economic necessities that farming had wrought. Because once men had land to protect and call their own, they needed children to help farm this land, and children to inherit it. And the only way a man in Neolithic Mesopotamia could properly identify his offspring was to mate with a certified virgin and call her child his own. It was the plow, then, that gave birth to marriage as we know it, allowing men across the early agricultural world to protect their property by controlling their wives' fertility.

The book turns next to the steam-driven changes of the Industrial Revolution, which slashed across the Western world in the eighteenth century and transformed everything in its wake. This revolution was mostly about machinery, about engines and railroads and looms. But as these inventions transformed the face of industry—indeed, as they *created* what we now think of as industry—they also transformed the shape and concept of the family. Men became industrial workers, tied to the factories and machines that gave the age its name. Women became the keepers of home and hearth, tied equally to their offspring and their wage-earning men. The stay-at-home wife, the breadwinning man, the housewife, the laborer, the salary man, the little woman—these are all creations of the Industrial Revolution, forged in the novel furnaces of steam and coal and steel.

We enter, then, the twentieth century, when innovation broadened and its pace accelerated. This was the era that gave us cars and refrigerators and washing machines—industrial machinery reconfigured for home use—followed by an avalanche of electronics and pharmaceuticals. Radios. Televisions. Antibiotics and contraceptives. From their inception, these products focused on the home and the individual, pushing innovation into the very core of people's lives. Automobiles gave humble farmers and housewives the dream of mobility, both geographic and social. Household appliances dramatically reduced the burden of housekeeping, freeing women from the daily drudgeries of cooking and cleaning and maintaining a home. And contraceptives separated sex from reproduction, allowing both men and women to control what had heretofore been left largely to nature. Feminism is a direct result of these technological shifts, a massive social movement ignited and enabled by innovation.

Fast-forward again to our own era and to a future whose technical prophecies are already becoming real. What happens to our notions of marriage and parenthood as reproductive technologies increasingly allow for newfangled ways of creating babies? What happens to our understanding of gender as medical advances enable individuals to transition from one set of sexual characteristics to another, or to remain happily perched in between? What happens to love and sex and romance as we migrate our relationships from the real world to the Internet? Can people fall in love with robots? Will they? And how will our most basic notions of humanity themselves evolve as we entangle our lives and emotions with the machines we have created?

To help answer these questions, *Work Mate Marry Love* draws upon an unlikely body of theory: Marxism. Or, more precisely, the study of historical materialism that Karl Marx developed in the late nineteenth century along with Friedrich Engels. Let me pause on that for a moment and elaborate. I am not making an argument here in defense of Marx's theories of politics, or even his economics. I am certainly not defending Marxism as it has been interpreted and implemented by various twentieth-century Communist regimes. Instead, I am arguing that, as a historian of technology and technological change, Marx offers insights that remain surprisingly valid today. He, along with Engels, was one of the few theorists—of his generation or ours—to link technological change with its societal implications, to connect the personal with the political.[5]

One needs to piece together these insights carefully from the massive body of work that Marx and Engels created, especially since the link between technology and social change was rarely the focus of their thought. But their foundational argument is that society and social structures are prodded always by the material circumstances that define them, and that these circumstances, in turn, are the products of technology and technological change.

Now, take that insight and apply it to some of today's fastest-breaking technologies—artificial intelligence (AI), for example, or in vitro fertilization (IVF). In the first case, faster and more powerful computers are increasingly able to do things and learn things that were once the sole province of humans. Machine tools equipped with even rudimentary bits of artificial intelligence can displace manual laborers on the factory floor; cars enhanced with AI-enabled navigational systems can and will supplant human drivers. But the change doesn't stop there. Instead, as waves of innovation surge across a society's shop floors and superhighways, they will wash inevitably into its kitchens and bedrooms, its families and romances, as well. Because the erstwhile laborers, having been replaced by machines, will need to come home. They will need different ways of constructing an identity and an income, different ways of spending their time and structuring their relationships. A change in the means of production thus ripples across the broader society, and into the nooks and crannies of life.

The second case—IVF—is even more dramatic. Here, technology is upending the most basic production process of all, giving humans an array of new ways to conceive and create a child. Already, would-be parents can select from a growing buffet of eggs and sperm and wombs. They can review an embryo's genetic composition and request specific traits—a girl or a boy, for example, with blue eyes or brown. Soon, scientists will be able to pinpoint these attributes with great precision; in the not-too-distant future, they may even be able to shape both egg and sperm from a single individual, giving people in effect the means to recreate themselves. Such possibilities are already big news in the relatively cloistered field of assisted reproduction. What I want to suggest is that they are bigger still—big enough, in fact, to reshape not only the families of children born from these emerging technologies but also the very notion of families themselves. Because if the nuclear family is essentially a means for two people to conceive, protect, and support a child, and if that family has been built around the two people—one fertile man and one fertile woman—needed to produce that child, then changing this most basic means of production will inevitably change the structure of the family as well.

The line between technological innovation and social change is rarely linear, or even clear. Instead, technology bursts forth while societies stumble and adapt. The inventors—from Henry Ford to Mark Zuckerberg and Elon Musk—are lauded or pilloried; the recipients are lost to history, often seeing no connection between the technologies that define their era and the personal decisions they either make or have thrust upon them. Today, millions of people are falling in love (or at least having sex) with partners they never would have encountered in the pre-Internet age. Most of them, presumably, don't see their dates and romances as linked, fundamentally, to technological change. But they are. And the families they form, the children they eventually produce, will be creations of technology as well—more interracial than their parents, more genetically dispersed, changed at the most fundamental level by inventions that will quickly feel as humdrum as the tea dances and church socials they displaced.

MOST OF US remember when technology was born. Or, more precisely, when it burst into the corners of our own lives, shaking the foundations of what once had been mundane. There was life as we had known it, simple and unchanging, and then something shattering, wondrous and new, that changed the contours of what we knew and how we would behave. For my Greek mother-in-law, born on an Aegean island in 1929, it was the arrival of the passenger car. For me, it was the personal computer. For my children, conceived at the dawn of the digital age, it was a stream of inventions, ticked off in sequence by the moment of their teenage years. Video games. Smartphones. Social media. Someone once said, and I believe, that the definition of "technology" is all the things that were invented after you came of age. Everything else—the car for me, computers for my children—is simply life as you've always known it.

Once these inventions crash into our lives, though, we tend to embrace them at lightning speed. It took only around fifty years, for instance, for electric lamps to morph from experimental curiosity to household necessity. Computers conquered the world between 1946, when ENIAC, the world's first general-purpose digital computer, was unveiled, and 1981, when IBM introduced its first personal computer. And the iPhone, hatched in 2007, had blanketed the globe by 2017 with more than a billion units, each snuggled into a palm as if it had been there forever. Even when we're scared by technology's arc, when we fear the implications of what a new world might bring, we seem to race forward nevertheless. Into thickets of nuclear weapons and robotic drones and the unfolding possibilities of artificial intelligence.

When technology moves as quickly as it is moving right now, though, it is crucial to remember just how wide its impact is, and how intimate. Because the products being crafted in the laboratories of the future are not destined to remain in those laboratories forever, or even for very long. Instead, like every breakthrough invention that preceded them, they will quickly break into our own lives as well, shifting the contours of how we live and love and play. In the past, the prevailing modes of production shaped a world dominated by heterosexual, mostly monogamous, two-parent families. In the future, these patterns are almost certain to regroup, creating entirely new norms for sex and romance, for the construction of families and the rearing of babies. More specifically, over the next few decades, heterosexual monogamy as we know it is likely to become increasingly antiquated, relegated by technological change to being just one possible structure among many. Gender—that most basic divide between girls and boys, women and men—will blur and expand, reshaping itself into a kaleidoscope of shifting identities. Reproduction will become more conscious and constructed, a way for individuals (and particularly the rich) to plot both their children's future and their own legacies. And, most dramatically, perhaps we will fall in love with nonhuman beings and find ways to extend our human lives into something that begins to approximate forever.

It is always dangerous to peer too closely into the future. Like Albert in the ecstasy of his Crystal Palace, we are likely to get much of it wrong, to fixate on the playthings that obsess us at the moment rather than the seismic shifts rumbling more subtly below. What Albert got right, though, and what *Work Mate Marry Love* seeks to explore, is the strong line that connects even the most industrial of technologies to the most intimate corners of our lives. Plows to monogamy. Dishwashers to feminism. And robots, AI, and assisted reproduction to whatever is about to come next.

As human beings, we seem preternaturally destined to build machines. Burly, smart, shiny machines that capture our vision and fantasy of the future. We build those machines to exact some greater measure of control over our lives—to work faster or more efficiently, to extract resources or cure disease, to unearth information and trumpet our tales. But the arc of technology doesn't end there. It never has, and never will. Instead, the machines we create begin to re-create us as well, to change the work we do and the lives we lead and what we define as good.

As technology evolves, in other words, so do we.

A FEW CAVEATS are in order before we begin. First, while this book is wide in scope, its review of technology is by no means exhaustive. I have focused on what I deem to be the most revolutionary inventions of our past, present, and near-term future, but other authors might well have chosen a different list. Second, although the book aims to describe causal relationships that are universal in their effect, the narrative is clustered around those people and geographies that have tended to be at the forefront of technological change. It is therefore perhaps unduly concentrated on the Western world, and on those individuals—generally white, educated, eventually wealthy men—whom history records as our greatest inventors. Finally, *Work Mate Marry Love* is not a book of scientific proof. I am not aiming to convince you that every connection I describe played out exactly as I suggest, or that historical events could not be viewed through a very different set of lenses. Rather, my objective is to use the patterns of the past to sketch a template for the future that is coming fast upon us, to construct a narrative that provides some guidance as to what we should embrace about this impeding future, what we should fear, and what, if anything, we can do to shape and constrain it.

My aim, in other words, is ultimately to tell you a convincing story about how our future—and that of our children and grandchildren—is likely to unfold.

But to do that, I need to begin with the very ancient past.

Part I

The

Way

We

Lived

1.

Life Before
the Machines

**The human career divides in two: everything before the
Neolithic Revolution and everything after it.**

—Ronald Wright, 2004[1]

**To the woman He said: "I will greatly multiply your pain in
childbearing; in pain you shall bring forth children, yet your
desire shall be for your husband, and he shall rule over you."**

—Genesis 3:16

We don't know much about how our ancient ancestors lived and loved.
Perhaps their romantic lives were as tumultuous as ours, filled with un-
requited yearnings and torrid affairs. Perhaps they loved their offspring
with the same ardor, cooing over their newborns and mourning a
child's untimely death. We just can't say.

What we can at least surmise, however, is that it was tools that
laid the foundation for what we now call civilization. For hundreds of
thousands of years, stretching from roughly 500,000 B.C. until around

8000 B.C., our prehistoric ancestors lived in small, nomadic tribes, clustered around the world's most fertile regions.[2] Their family lives were porous, their children were raised by bands of female relatives, and marriage as we know it did not exist. Food—seeds, stems, nuts and fruit, shellfish and small mammals—was foraged for and occasionally killed, and home was wherever the band settled for the evening. Perhaps life on this distant savanna was, as some anthropologists have argued, a slow-moving idyll: free love; free food; no pollution, guns, or traffic jams. Or maybe it was just nasty, brutish, and short. Again, we just don't know.

In either case, though, we know that life began to change—gradually at first, and then with an earth-shifting momentum—roughly ten thousand years ago, when bands of settlers around the Nile River and in the valley of the Tigris and Euphrates launched what would much later be known as the Neolithic or Agricultural Revolution—one of humankind's first and most far-reaching technological revolutions. Through accident or accumulated wisdom, small groups of what were once nomadic people learned to cultivate the land, rather than just living off its natural bounty. They learned to farm and to harvest, and to build the villages that could protect both themselves and their crops. In the process, as Marx pointed out, they also developed both an early form of private property and the need to protect it.

In particular, those who developed land and the tools to farm it wanted to keep that land and make it theirs. They no longer wanted to share the fruits of their labor with an entire band of comrades, but only with those closest to them: those who helped in the fields, and who would remain tied and committed to the land after their own deaths. In other words, they wanted to give their hard-won rewards to their children—which meant that they needed to know just who those children were. And thus, without much romance, the institutions of marriage were very likely formed. Men with tools and property married virginal women and demanded their fidelity for life. The children born of these unions were then presumed to be the tool wielder's rightful heirs—the descendants who would work his land and inherit his property.

What we think of, then, as love—or at least mating and till-death-do-us-part marriage—is actually the probable by-product of technological change. For millennia, men and women lived mostly in groups, spending and sharing their lives with fellow travelers rather than a single mate. Then along came the plow, and everything changed.

Love Among the Cavemen

According to the most recent archaeological evidence, *Homo sapiens*—our most direct and distinctly human ancestors—emerged on the planet sometime around two hundred thousand years ago, living first in the lush forests of East Africa and then spreading gradually to the east and north.[3] These early foremothers of ours had basic tools, the evidence suggests, and lived, as we've noted, in small communal bands. They were hunters and gatherers by occupation, living on the fruits and nuts they foraged from the lands around them, along with occasional game.[4]

Scholars of this period do not agree on how these early societies structured themselves. Some argue that the first *Homo sapiens* lived in female-dominated groups, comprised of mothers, sisters, and young children. Others believe the groups were primarily organized around males, while a third school argues for more gender mixing, with one dominant male consorting with several females and their children.[5] All of these theories, though, agree on a core and critical proposition: by around 70,000 B.C., when *Homo sapiens* started to become the dominant human species, our ancestors gathered and lived in nomadic groups.[6] There were no isolated couples living alone with their children, no permanent homes or belongings.[7] Instead, the dominant societal unit was the group, with food, children, and responsibilities shared across the community. There was little scope for human choice in these arrangements, shaped as they were by the imperatives for survival.

Within this structure, anthropologists again surmise, men and women divided certain labors among them. Men, blessed with greater height and physical strength, assumed the primary burden of tracking and killing large game. Women—with young children in tow—gathered plants and small animals, and manufactured whatever clothing and cooking implements could be made. The sexes were dependent upon each other, therefore, but it was a dependence, and a power relationship, that seems to have run in both directions. Women needed men to provide occasional bursts of high-value protein; men needed women to clothe and feed them on a more regular basis. Women needed men to create offspring (though they may not have realized this connection), and men needed women to nurse those children and bring them to adulthood.

Sexually, the bargain appears to have been a sort of open-ended bonding. Men and women mated primarily with a particular partner, and then stayed nearby to care for the children they bore. But partners moved on as their children matured, and they engaged more freely, albeit still clandestinely, with other members of the group.[8] Maybe there was love in those early days. Maybe even something that looked like fidelity or romance. But all we know—from both the archaeological record and suggestive anthropological studies of modern hunter-gatherer tribes—is that there was sex, and sharing, and offspring who managed to survive the struggle for selection.[9]

One of the first social scientists to draw wider-ranging conclusions from these early structures was Friedrich Engels, the pathbreaking German philosopher who, along with Karl Marx, had already written the magisterial *Communist Manifesto* and crafted the theoretical foundations of communism. So he had his opinions. But his conclusions about prehistoric life have stood the tests of both time and thousands of would-be-contrary scholars. Essentially and critically, Engels argued, prehistoric societies were small and collective, marked by a gentle division of labor that bound the members of the group to one another, but not to any specific or long-term pairings.[10] Men did more of the hunting; women more of the gathering, trapping, cooking, and small-scale cultivation. There was no private property during this long era, no marriage contracts or feminine submission.[11] Until around 8000 B.C., when the world's first technological revolution transformed everything that world had yet known.

Plow. Shares. Remaking the Ancient World

Standing as we do today in an era of artificial intelligence, space flights, and life-altering pharmaceuticals, we find it difficult to comprehend the power of the world's first technological revolution, or to understand how deeply society was shattered and then restructured by the invention of very basic tools. It's hard even to think of the plow and the hoe as technology, given their simple, almost intuitive shapes and functions. The hoe, after all, is little more than a handheld pole attached perpendicularly to a blade. The plow—as it appears in various incarnations

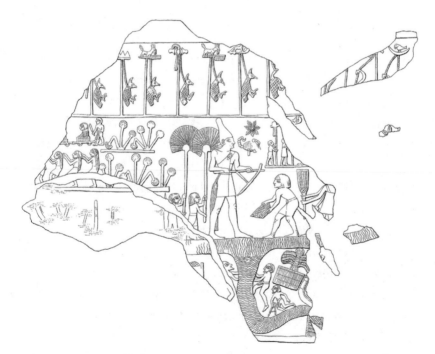

Hand plow from the late Predynastic to early Dynastic Period in Egypt (about 3100–3000 B.C.), from the mace-head of King Scorpion

around the world—is essentially a large digging stick, dragged by a pair of oxen or water buffalo.[12] Yet together they changed the world and the ways in which human beings lived in it.

We will never know who created these tools or how they eventually swept across the globe.[13] Between 9500 and 8500 B.C., however, agricultural practices in the hill country of southeastern Turkey, western Iran, and the eastern coast of the Mediterranean began to change, slowly at first, and then with a rapidly increasing velocity.[14] Humans, who by this point had lived as nomads for tens of thousands of years, started to become more sedentary, forsaking foraging for farming, and learning to domesticate both animals and crops. Initially, they focused on the resources around them—on the sheep and goats that already graced their hills, for example, and the olives and grapes that naturally grew there. As they began to understand the process of cultivation, though, the settlers developed tools to encourage and expand it: hoes, first, to stir up the soil and embed the seeds, then plows to create larger and more productive fields and irrigation channels to keep them watered. With these tools, settlers became farmers, growing crops such as wheat or rice consciously now, and with an efficiency that increased over time. This ability to *control* nature's bounty, rather than just searching for it, was the first major step that humans took toward technological mastery—the step that gave us as a species the unprecedented capacity to rule (and shape, and perhaps destroy) the planet. In the process, though, it also made us settle down. Because once Neolithic nomads learned to farm, and to depend on farming to provide the bulk of their foodstuffs, they had to stay by these farms, tending to the crops and watching them grow.

This was the move—dominated by agriculture, enabled by technology—that changed the very nature of human society. Because once the production of food was truly a *production*, it demanded a certain set of social and economic relationships around it. Most basically, it demanded permanent settlement—a dwelling for the farmer to live in, and a safe place in which to store his tools and harvest. Producing food also required a new and wider range of implements. Foragers need little more

Ancient Egyptian ard, or "scratch plow," circa 1200 B.C., from the burial chamber of Sennedjem

than a container in which to carry their haul. Hunters, even big-game hunters, need only a weapon and a means of cutting and cleaning their prey. Farmers, by comparison, need *stuff*. Not just hoes and plows but also blade sharpeners and draft animals, granaries and seed storage bins, mortars to crack open the tough grains and pots in which to cook them. No wonder, then, that archaeological remains from early agricultural societies—regardless of where they occur—are so rich in remnants. Potsherds, hoe blades, urns and water jugs and baskets—all became necessary to support an increasingly permanent, agriculturally dependent society.

This growth of goods, in turn, meant that people suddenly needed to protect their goods, to define what was "theirs" and keep it in their possession. Before this move, when life was communal and individuals generally lived on the food they could find with the tools they could carry, property made no sense. People had few goods; they moved from place to place. But once they settled down, once they needed things, once they started to *produce*, men and women needed to control their possessions and make them theirs. As the economic historians Douglass North and Robert Paul Thomas wrote in an influential 1977 essay, "It is inconceivable that, from the very beginning, the first farmers did not exclude outsiders from sharing the fruits of their labor."[15] And thus private property was eventually born.[16]

It's not intuitively obvious that gender roles would change as a result of technology and property, or that men and women in agricultural societies would rearrange their relations—economic, sexual, and familial—in such radical fashion. Yet they did, transforming life and love for the next ten thousand years.

The first and most obvious change was that people stopped moving. As they turned increasingly to planted crops for food, these first settlers abandoned the nomadic ways that had characterized their lives for so long. As a direct result of this physical settling down, men and women were put into direct proximity, with men, in particular, no longer called away for the large-game hunts that had once occupied at least a portion of their time.[17] Focused now on farming, men stayed closer to home and played a larger role in the day-to-day production of food. As agriculture grew more advanced and populations moved from hoes to plows, men's roles increased again: unlike hoes, which could be used by women or children to clear small plots, plows required significant physical strength both to drive a much larger piece of machinery and to handle a team of oxen, cows, or water buffalo. In societies that shifted to plow-based agriculture, therefore, women were gradually pushed out of the production process, losing over time the ability to support themselves, and the power that inherently comes with such independence. According to one recent economic study, even today, centuries after adopting this technology, women from plow-based cultures are less likely to participate in their society's labor force than are women from cultures that never embraced the plow.[18]

Likewise, archaeologists and economists have estimated that women living before the Agricultural Revolution provided more than half the calories their communities needed.[19] After the revolution, their contribution declined precipitously, and women were relegated largely to the secondary tasks that larger-scale farming now required: weeding and picking and grinding the grain. In the process, as Helen Fisher notes, "Women lost their ancient, honored roles as independent gatherers and providers . . . [and] women were judged inferior to men."[20]

Where women did become more valuable, though, was in the production of children. Of course, they had long played this role physically—nothing changed there. But the value of children changed in a world dominated by settled agriculture and private property. During the nomadic period, children imposed a tangible cost on the parents who conceived them. They had to be carried over long distances and for long periods of time—nearly five thousand miles over the first four years of life, according to one estimate; they weren't good at either hunting or gathering for many years; and their cries likely drew the unwanted attention of predators.[21] Accordingly, women living before the Neolithic Revolution bore relatively few children. They nursed their babies for extended periods (which reduced the chance of conceiving); they killed or neglected infants they could not afford to support; and they probably relied on plant-based abortifacients to prevent unwanted pregnancies.[22]

As once-nomadic tribes became more sedentary, however, the calculus around both women and children began to change.[23] Most obviously, children didn't need to be carried long distances anymore, and could easily be taught to tend to fields and livestock. In economic terms, they became less costly to the community, and potentially more valuable. Moreover, as these communities increasingly forsook their hunting and gathering in favor of cultivation, they became dependent on agricultural yields: they *needed* the crops to grow and the harvests to be bountiful. Which meant, in turn, that they needed labor—as many hands as possible to tend and till the fields.

One way of obtaining this labor, as many historians have ruefully noted, was by capturing other people and forcing them to work. This is why the earliest powers of the Neolithic Age relied heavily on slaves.

They taxed the farmers within their borders and captured surrounding peoples to build walls, dig canals, and shoulder some of the physical drudgery that increasingly large-scale agriculture demanded.[24] The other way of generating labor, though, was through birth. Specifically, by prizing and prioritizing reproduction, early agricultural societies could generate the labor pool upon which they now depended; they could create what was arguably their most important resource.[25] But doing so meant shifting women's roles, prodding them away from the shared work of production that had been their fate until this point and into the more cloistered and confined work of *re*production. And thus women's worth during the Neolithic Revolution—and for millennia to follow—became deeply and inextricably intertwined with their ability to breed. As James C. Scott concludes in his history of the earliest agricultural states: "A combination of property in land, the patriarchal family, the division of labor within the domus [residence], and the state's overriding interest in maximizing its population has the effect of domesticating women's reproduction."[26]

What made this differentiation even starker was the inevitable coupling of sex, property, and progeny. Because once children had real value as both labor and heirs, and once men had invested time and energy in cultivating what was now their own land, it became critical for them to know which children were theirs. And long before genetic testing, the only way to ensure paternity was to control fertility, binding an individual woman to "her" man, and preventing her from having sex with any other.

As the Neolithic Revolution unfolded, different societies solved the paternity problem in a range of ways, from monogamy to polygamy, harems, bride prices, and concubines. Nearly without exception, though, their solutions involved some combination of marriage and fertility controls; some way of guaranteeing that women had sex with one man, and one man only. Fidelity therefore (at least for women) became a prized societal value, and marriage a dominant social structure. In describing this crucial shift, Helen Fisher writes: "With plow agriculture came general female subordination, setting in motion the entire panorama of agrarian sexual and social life, including the rise of the sexual double standard."[27] Casper Hansen, Peter Jensen, and Christian Skovsgaard, a trio of Danish economists, are blunter: "Patriarchy," they conclude, "has its origin in the Neolithic Revolution."[28]

And so does marriage—particularly the monogamous, virginal, and formally sanctioned kind. During the latter portion of the Paleolithic Age, men and women may well have clustered into couples; they may have stayed together as a pair for some extended periods of time. But there is little evidence to suggest that these pairs were bound to each other in a permanent way, or that their sexual lives were unduly restricted.[29] Once agricultural technologies unleashed their revolution, though, and property became private, marriage became an increasingly crucial institution. Because it was through marriage that men could formally claim their wives and control their sexuality.

Accordingly, evidence from across Eurasia bears witness to the increased importance of marriage ceremonies and marital customs beginning around 3500 B.C. In Sumer, for example, laws began to codify what constituted a valid marriage—evidence of certain kinds of marriage gifts, usually, and a dowry. In Ur, an important Mesopotamian city-state, legal codes similarly began to define both polygamy and divorce. And across Mesopotamia and much of the Near East, recovered texts from the period suggest, marriage rites started to emerge, often involving some kind of sacred ceremony, a bathing ritual, and the exchange of gifts.[30] The groom gave his bride's father a payment known as *terhatum*—often viewed and legally interpreted as "the price of a virgin."[31]

To be sure, the monogamy of this period ran almost always in only one direction.[32] Men throughout the Neolithic world were free to take as many wives (or concubines) as they could afford, and as societies grew wealthier, having more wives became both a symbol of riches and a way of expanding them: more wives meant more children and thus labor for the land. King Solomon, ruler of the Israelites in the tenth century B.C., supposedly had seven hundred wives and three hundred concubines. Amenhotep III of Egypt had a harem that included two Syrian princesses, two Babylonian princesses, "droves" of Egyptian women, and one Great Wife.[33] Women, by contrast, who bore the heirs, had to be strictly controlled, "lest," as one ancient Sanskrit text warned husbands, "the seed of others be sown on your soil."[34] Accordingly, women's bodies came increasingly to be regarded as the property of their fathers and husbands, valuable commodities to be preserved, bartered, and used by the men in their families.[35]

The ancient civilization of Sumer offers one compelling glimpse into how these transformations played out over time. Located in modern-day Iraq, Sumer rose to prominence around 3000 B.C., when an accumulation of silt carried downstream by the Tigris and Euphrates Rivers built a vast and abundant plain, the heart of what would become known as the Fertile Crescent. Over the next one thousand years, an emerging class of priests and kings built an astonishing network of canals, dikes, and dams, controlling the waters that simultaneously nourished the landscape and threatened it. As the fields became increasingly productive, yielding predictable harvests each year of wheat and barley, the region's population swelled and a growing class of people—mostly men—was able to turn its energies and attention to pursuits other than farming. Sumerians developed writing tablets and wheeled chariots, abacuses, musical lyres, and ornate furniture—the first recorded civilization to have developed such a wide range of products beyond those necessary for subsistence agriculture. At its height, around 2000 B.C., Sumer boasted walled cities filled with temples and lush fields, and a population that may have topped 1 million.[36]

Yet wealth, as is so often the case, also took its toll. Sumer's riches allowed its leaders to develop wide-ranging authoritarian powers, using conscripted labor to build their temples and waterworks, for example, and declaring all inhabitants of their lands subject to the king's rules and taxes. Women were treated particularly harshly in this burgeoning civilization, subject to the world's first (or at least earliest recorded) set of gender-specific rules. Starting around 2300 B.C., female adultery became a crime punishable by stoning and a woman found guilty of verbally disrespecting a man was to have her "teeth crushed by burnt bricks on which her guilt had been inscribed."[37] Some five centuries later, King Hammurabi's famous code essentially treats women as possessions, offered by their fathers for a price and legally expected to provide their husbands with children, sex, and lifelong fidelity. Under Hammurabi's reign, a wife who was found "surprised"—that is, having sex—with another man was to be tied and thrown into the river.[38] Men, by comparison, were free to elope, claim concubines, and rape girls or women so long as they weren't already betrothed to other men.[39]

History records these shifts most poignantly, perhaps, in the faces of its gods. Before the Agricultural Revolution, hunting and gathering cultures seem to have worshipped a panoply of mostly female figures— earth mothers, fertility figures, goddesses. We know of these goddesses from the fragments that cling still to ancient histories—Isis, Ishtar, Freya, Gaia—and can see their remnants in a trove of totems unearthed across Europe and Asia that archaeologists collectively call "Venuses." Although these figures vary widely in terms of material and origin, they share a striking similarity. All, as the photos on page 30 demonstrate, are clearly and exuberantly female, with oversized breasts and hips; all, the archaeologists surmise, were used as items of worship.[40]

As societies moved to agriculture, however, their gods became male. In Mesopotamia, for example, the adoption of the plow around 4000 B.C. corresponded to a shift away from the region's traditional adoration of mother-goddesses and priestesses, and toward almost entirely male gods.[41] In Europe, phallic symbols first emerged among late Neolithic cultures.[42] And when the early Hebrews entered Canaan, they followed Moses's lead in abandoning their worship of a panoply of gods and goddesses and pledged their sole allegiance to Yahweh, the Lord God.[43] The allegorical interpretation of Genesis is therefore intriguing to contemplate: the Israelites chose their God, and Eve was banished from Eden, condemned forevermore to obey her husband and labor in pain.

By around 2500 B.C., then, in cultures and civilizations that stretched from the Yangtze to the Nile, men and women were increasingly

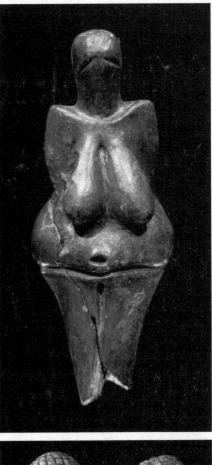

Clockwise from left: The Venus of
Willendorf (Austria, 24,000 B.C.), the
Venus of Dolní Věstonice (Czech
Republic, 28,000 B.C.), and the
Venus of Brassempouy (France,
22,000 B.C.)

arranged in fixed and formal pairs. Having become dependent upon cultivated crops such as wheat, corn, and barley, humans were settlers now, living in camps and villages scattered across the world's most temperate regions. The rough equality that had characterized their forebears was gone, as was the easy promiscuity that had once prevailed. Men and women were marked forever by their gender, and divided into separate spheres. Where the sexes came and stayed together was in marriage—a formal state now, sanctioned by the spouses' families and whatever kings and gods they obeyed.[44]

From time to time, historians and anthropologists wonder why men and women chose to change.[45] Why would women have ever given up the freedoms that once were theirs? Why would either gender abandon the nomadic life, filled as it apparently was with a diversity of foodstuffs, for a backbreaking existence marked by driving oxen and eating grain? Why, as even the Bible seems to question, did we ever want to leave the garden that was Eden?

Some suggest that it was climate change, a long, slow warming-up and drying-out that transformed the once-lush marshlands of the Middle East into more arid and unforgiving terrain.[46] Others blame the power-hungry advances of early Neolithic rulers, who enslaved the region's nomadic tribes and forced them to both settle and produce.[47] And another group points to a dizzying rise in global population (from an estimated 5 million in 5000 B.C. to 170 million by the year A.D. 1) that would have sharply constrained the natural resources available to hunting and gathering tribes, and pushed them into cultivation instead.[48]

In any of these narratives, though, the overarching answer is that we didn't actively choose—no more than any humans ever choose either to innovate or to be affected by invention. Rather, just as the Victorians would harness steam, just as a handful of disaffected college students tinkered in their dorm rooms and garages and created what would become computers and social media, our ancestors stumbled into agriculture and all that it became.

Technology had befallen the species and changed it forever.

One by one, the civilizations of the ancient world took the technologies of the Agricultural Revolution and made them flourish. Driven by plows and irrigation technologies and a host of related tools that followed in their wake, agricultural innovation spread across the great empires of the Near and Middle East: Mesopotamia, Egypt, Greece, and Rome. It transformed large swaths of China and cropped up more patchily in the South American empires of the Incas, Aztecs, and Maya.[49] Where technology stayed dormant—such as in Australia and parts of sub-Saharan Africa—populations remained mostly small and nomadic.[50] But where technology wandered, it stayed, establishing or intensifying some of our most familiar social institutions: Marriage. Family. Monogamy. Wealth. And power.

The subsequent advances of this age lay mostly with improving agricultural techniques and building infrastructures, both physical and political, to support them. Particularly important were technologies for moving water—bringing it from its sources to the fields and villages that were now locked in place; storing it; and ensuring its supply in drier times. In Mesopotamia, for example, local rulers organized crews to drain the marshes between the Tigris and Euphrates Rivers, and to construct the canals and irrigation systems that kept the plains fertile. In Egypt, pharaohs organized massive efforts to track and control the Nile's annual flooding. And in China, Yu the Great became legendary around 2200 B.C. as the "ruler who controlled the waters," dredging a network of sophisticated canals to irrigate settlements along the Yellow River.[51] In all these instances, water became far more than a way to quench a growing population's thirst.[52] It was the source of economic growth, and of political power. By controlling his people's water supply, a leader effectively controlled those people as well. Maintaining a constant flow of water—through the cisterns of the Assyrian empire, the dikes of early Sumer, or the aqueducts of Roman Europe—entailed raising funds and troops and supplies to build them. Turning off the water meant death.

Combined with the wealth and population expansion that larger-scale agriculture generated, these water-induced power structures meant that the growing societies of classical antiquity also witnessed growing inequalities among their people.[53] City-states and empires were increasingly divided into various classes: free and slave, patrician and plebeian, ruler and ruled. The most basic division, and one that became ever more rigidly fixed during this era, was between man and woman, with women across nearly all of the ancient world being forced into a state of formal submission. In Rome, for example, the head of every household was the *paterfamilias*—father, son, uncle, or grandfather, he was always male, and the effective ruler of all women and slaves living within his domain.[54] In Athens, similarly, girls were married off as teenagers and a wife's infidelity was grounds for divorce.[55] To enforce this role, Athenian women were generally confined to separate quarters and guarded, Aristophanes reports, by dogs trained to scare off adulterers.[56]

With these newfound social structures thus firmly in place, the great civilizations of the ancient world developed the wonders for which history best remembers them: the Great Pyramid of Giza, the Colossus of Rhodes, the Lighthouse of Alexandria. Less celebrated, though no less wondrous, were the technological marvels that developed during this time: iron tools, the potter's wheel, and, as Engels writes, "the beginnings of architecture as art."[57] The revolution, though, had already taken place, launched with the knowledge of how to plant seeds, and transformed through the manufacture of hoes and plows and dams and aqueducts—the means through which humans began to change nature, and then themselves.

Love Among the Ruins

After the passing of the great classical civilizations, after Alexandria burned and Athens fell and Rome dissolved, the pace of innovation slowed. There were bursts of genius, of course, over the next two thousand years, and inventions that gradually accreted toward progress. But the revolution was over, as were the seismic social changes that had accompanied it. Men and women; rulers and ruled; husbands, wives, and children settled down into what became wholly familiar patterns. So familiar, in fact, that nearly everyone who lived under them presumed them to be simply the way of nature, and the will of their newly created gods.

At the core of this system, across Europe and Asia and most of the so-called civilized world, was the institution of marriage—a lifelong, heterosexual union endorsed by church and state. A woman's virginity was heavily protected before marriage (through a combination of tradition, inspection, and, occasionally, brute mechanical force) and female adultery was considered a crime.[58] Children were valued as laborers and heirs, and families, for the most part, stayed together. Throughout the long era that stretched between the Roman Empire and the Industrial Revolution, most of the world's population lived in villages or small towns, farming—wheat, or rye, or rice—and tending to small flocks of livestock.[59] People worked all the time, waking with the sun to feed the chickens or tend the fire.[60]

Except, of course, for small populations of royalty and aristocrats, who generally rose to prominence through war, and stayed in power by compelling the residents of their lands to furnish them with both taxes and troops. When it came to marriage, though, the elite of the Middle Ages behaved surprisingly like the serfs. They married overwhelmingly for practical reasons (to cement an alliance or grow a realm); they forbade adultery among women but ignored it for men; and they stayed together (in theory, at least) until death.[61] By the Carolingian era (the eighth through tenth centuries), as one history recounts, nobles "married one wife at a time, content to season their monogamy with concubinage."[62]

What held society in place throughout this long era was religion—faith, increasingly, in a single god whose word was law and whose powers were infinite.[63] In Europe, after the decline of the Roman Empire around A.D. 500, the Catholic Church slipped into the void of power, ultimately presiding over the Holy Roman Empire.[64] In the Middle East, the spread of Islam after Muhammad's death in A.D. 632 created millions of religious adherents, former atheists or polytheists who now believed in the power of a single god—Allah—and the importance of conforming to his laws. The growing power of religion worked to slow the pace of technological change across Europe and Asia. It also reified existing social structures, exhorting believers around the world to celebrate what was already theirs. Islam's central tenets, for instance, were simple and straightforward: "I fast and I eat," Muhammad enjoined his followers. "I keep vigil and I sleep and I am married."[65] Christianity similarly extolled the stability of its adherents' lives, urging them to focus their energies on God and prayer.[66]

For families, the impact of these doctrines was clear and all-consuming, detailing precisely what family life should look like and how aberrations would be treated. Marriage was now firmly fixed in both law and custom, defined as an act of God that "no man" should ever "put asunder."[67] Virginity was strictly prescribed for would-be brides, and fidelity after marriage was the rule: "The wife hath not power of her own body," warns the New Testament, "but the husband: and likewise also the husband hath not power of his own body, but the wife."[68] Sex was strictly for procreation, with children held out as gifts from God. Adultery was a sin for both parties, although the punishments meted out to women were considerably worse.[69] In most parts of the world, land passed from father to son, underscoring the value of paternity and the importance of producing male heirs.[70]

Looking back upon these practices with the gauzy eye of history, we tend, typically, to describe them as rituals of a particular age, traditions that arose and were reinforced through a combination of religious teaching and social norms. But it's much more than that. Because religion doesn't spring from a void; instead, like all social structures, it is very much a product of the age that produced it, and of the technologies that define that age. How else to explain the adoration of virginity across nearly all major religions and societies during this time? Or the fervent embrace of monogamous marriage—something that wasn't practiced in ancient times, and is fast falling out of favor today? No, this long era worshipped marriage and virginity because technology demanded it.

At a time when wealth was linked primarily to land, and land secured through still-nascent systems of property rights, marriage was the contract that provided security. Especially for peasant farmers, whose livelihood depended on their fields and families, any rupture of this structure—dividing the lands, for example, or losing a critical pair of hands—would have been disastrous. Farmers needed their wives, in other words, not so much for love as for money; they needed them to milk the cows and brew the beer and produce the children who would serve as farmhands. And, with few other options for their own economic survival, farmers' wives needed their husbands just as much. So the religious doctrines that prevailed during this time—sanctifying marriage, condemning divorce, celebrating procreation under wedlock—aligned precisely with the era's economic needs.[71]

This link was even stronger around the crucial question of paternity. So long as land was the primary source of income, determining who would inherit this land was understandably a major concern. Following the pattern that had been set in the early days of the Agricultural Revolution, most societies during the Middle Ages adopted customs or laws that gave inheritance rights to a man's legitimate sons, or even to his eldest son alone. Technologically, this preference made sense: it concentrated property rights in a relatively limited set of hands, ensuring that an individual's investment in his land would not be lost to neglect or competing claims. Societally, though, it imposed a fairly rigid set of obligations. First, men had to know—without question or doubt—who their children were. Second, they had to differentiate between legitimate and illegitimate heirs. And third, everyone around them had to agree to these distinctions.

Once again, marriage made this system work. In particular, marriage solved the problem of paternity so long as women were certifiably virgins at the moment they married and stayed forever loyal to their spouse. All children of a man's legal wife, therefore, were his. If he conceived other children—either before marriage or through the occasional dalliance—they could be ignored or conveniently disregarded as illegitimate. And if a woman produced children outside of marriage, she could be divorced or abandoned or killed.

IT'S HARD TO imagine that anyone living through this long era—be they nun or knight, vassal or king—would attribute the source of their social context to technological change. We rarely do. But the central institutions of family life that became embedded across most of Europe and Asia between roughly 3000 B.C. and A.D. 1800 were not preordained by nature or fate. They were neither permanent nor universal. Instead, they were the product of a specific moment in time, and of the technologies that prevailed. Marriage. Monogamy. Female subservience and fidelity. All had slowly but surely emerged from the cauldron of the Neolithic Revolution and its technological turn to plow-based agriculture.

Many other crucial technologies arose over this period as well. Stirrups, for instance, which massively increased the capabilities of a horse and its rider. Hand looms enhanced by treadles and spinning wheels. And windmills, created in the twelfth century, which cracked open a potential new means of power.[72] All of these, and many more, reshaped the world in their own ways, and affected the lives of those living through and after their development. But given the already-vast scope of this book's narrative, my goal is not to explore each major invention, or even to trace the broadest contours of technology's arc over this long period. Instead, I want to focus solely on the most seismic of shifts, on those that, like today's advent of digital technologies, promise to reshape and rejigger our lives in the most fundamental of ways.

The first of these great revolutions was humankind's adoption of agriculture.*

The second was our embrace of steam-powered industry, to which we now turn.

---

* One can count the discovery of fire as the very first technological revolution. Because fire was technically discovered and not invented, though, and because its history is nearly impossible to trace, I have followed the common practice of labeling the Neolithic Revolution as humankind's first.

2

Steam Heat

How the Industrial Revolution
Transformed Women, Men, and Work

All was expectancy. Changes were coming. Things were going to
happen, nobody could guess what.

—Lucy Larcom, *A New England Girlhood*, 1889

Ironically, one of the biggest shifts in the history of human invention
came from a humble and well-known technology. It was steam—simple
water heated into vapor—that catalyzed the Industrial Revolution and
subsequently transformed not only the worlds of commerce and finance
but the core of family and social relationships as well. Before the revolution, manufactured goods were made largely by hand, and often at
home; afterward, production moved into the factory and machines

invaded the home. Before the revolution, humans moved largely at the speed of their own steps, accelerated, at most, by horse-drawn vehicles or wind-blown sails.[1] After the revolution, they could almost literally fly, racing across miles and miles of terrain in steam-powered trains and ships. Steam built our modern economy and shaped, for centuries, the implications of what it meant to be a worker, a husband, and a wife.

Launched roughly around 1760, with James Watt's development of the steam engine, the Industrial Revolution stretched until the latter portion of the nineteenth century, ranging over that period to embrace ever-larger geographies and an ever-expanding network of innovation. As its name implies, it was, for the most part, a revolution centered on machines—big, muscular machines that employed evolving technologies of steam-powered automation to create entirely new modes of industry, transportation, and commerce. As the tools of the revolution expanded, the manufacture of goods moved from small-scale workshops to factory assembly lines. Rather than being hauled by cart or wagon, they were increasingly moved by the power of steam—on railways or steamships—and across vastly wider networks of trade. Financial information, once relayed solely by letter or newspaper, was sent via telegraph. Steel manufacturing and oil production both came, suddenly and massively, to dominate the industrial landscape. And the world of the Middle Ages—the world whose social structures had been roughly the same for thousands of years—was shattered. Or as Marx and Engels, whose thinking was a direct response to these changes, wrote: "In acquiring new productive forces men change their mode of production; and in changing their production, in changing the way of earning their living, they change all their social relations. The hand-mill gives you society with the feudal lord; the steam-mill, society with the industrial capitalist."[2]

At the same time, though, and more subtly, the Industrial Revolution also occasioned a revolution at home, changing the lives of those who lived during its advance and those who would follow in its tumultuous wake. Because once steam quickened the pace of both production and transportation, streams of people began to leave their physical households to labor away, working on large-scale enterprises—railroads, steamships, factories—and earning cash money in return. By the turn of the twentieth century, the bulk of workers across the now-industrialized world were employed outside of agriculture. And they were overwhelmingly male—a growing, increasingly urban workforce who left their wives at home.

This was the great social shift brought on by the machine age, a transformation that fashioned a new division of labor between men and women; between those who went out to work, earning an income to support their families, and those who stayed at home, tending to the tasks for which there was no pay. And thus the same kind of division of labor that was occurring in the factories of the industrial age—workers in one slot, owners in another—was replicated at home. Men in one role. Women and children in another.

We are living still in the shadow of the Industrial Revolution, lumbering under the industrial structures it smashed into place and the social orderings that developed as a result. The environmental devastation that confronts our planet was first ignited by the coal-fired factories of the late eighteenth and nineteenth centuries. So, too, was modern capitalism, which emerged from the capital intensity of these factories and the financial rewards they bestowed upon their owners. Wage labor as we know it arose during the Industrial Revolution, as did the modern state. These causal stories are well-known.[3] What is less often detailed, though, is the extent to which the Industrial Revolution also caused a domestic revolution. It was a quieter revolution, free from the clanging gears and fiery furnaces of its more tangible counterpart. It occurred away from the factories themselves, in the more intimate realms of the home. But as the Industrial Revolution spread around the world, it generated social rules and positions that were every bit as novel as railways and steamships. It created wage laborers, primarily male, and housewives, who were idolized as submissive and self-sacrificing. It created children who were adored and mothers raised to nurture. It sanctified marriage as an idealized partnership, with property, offspring, and housewares its desired output.

The revolution began in England, in the northern hills surrounding what was once the sleepy town of Manchester. It percolated there for several decades, as the inventions that would turn steam to power spread across various proto-industrial sectors. Textiles. Mining. Iron. Then, like steam itself, the revolution spread, and proceeded to conquer the world.

Seeds of Change: Origins of the Revolution

As had been the case with its Neolithic predecessor, the Industrial Revolution happened very slowly and then all at once. Some historians trace its origins to the accumulation of mechanical precursors that had evolved during the long Middle Ages.[4] Others cite the creation of financial instruments that enabled capital to accumulate and flow more freely, or the emergence of rational and humanist thought that characterized the Enlightenment.[5] All these advances were undoubtedly important. But the most direct antecedents emerged sometime around 1750, when potatoes and cotton became the revolution's two most literal seeds of change.

Prior to this point, European farmers had generally viewed potatoes with a combination of suspicion and disdain. Potatoes grew underground; they were dirty and stubby, and came from lands that most Europeans regarded as primitive. Potatoes needed to be boiled or baked to be edible, and weren't easily transformed into the bread or porridge that composed the mainstays of most European diets. But as European peasants started to experiment with potatoes in the early eighteenth century, their advantages became obvious.[6] Potatoes were easier to grow and harvest than were grains like oats and barley. They produced more calories from smaller and less fertile plots of land. They were tastier, too, and generally more resistant to disease.[7]

As potato farming spread across northern Europe, the population of these areas grew accordingly. Britain's population, for example, doubled between 1750 and 1800; France's rose by 50 percent over that same period. In Holland and elsewhere, farmers took advantage of increased agricultural yields and began to convert former swamplands into fields, rotating their plantings between potatoes and other crops, and generating higher overall output as a result.[8] As these developments played out over time, Europe grew richer and more populous, generating what Marx would subsequently label as surplus capital and a surplus source of labor.[9] These would provide the basis for the revolution to come.

Meanwhile, an expanding population had also come to embrace a newly available product: cotton. For millennia, most of the clothing worn by Europeans was made of linen or wool. These were fibers that could be produced largely at home. Peasant farmers would typically either grow small fields of flax or keep flocks of sheep; their wives would spin the yarn, and then weave it into cloth on rough-hewn wooden looms. Most people didn't have more than a few items of clothing, and they didn't wash them very often.

Cotton didn't appear on European markets until relatively late. Although it had been harvested and woven for thousands of years, it was a southern crop, grown primarily in the warmer lands of India, eastern Africa, and Central America. Cotton cloth from India was traded across the Roman Empire and South Asia as early as the sixth century, but it was expensive and bulky, and the Arab merchants who specialized in its trade only ventured slowly beyond the Mediterranean.[10] As they did, though, cotton became a coveted luxury. Cotton cloth was easy to weave. It was easier to wash than linen or wool, and softer against the skin. It could be dyed an infinity of shades and sewn into an infinity of fashions. Cotton, as one nineteenth-century industrial report proclaimed, was "one of the many wonders of the world."[11]

Once again, it was production from the New World that jolted the Old. As settlers from Europe had slowly expanded across the Americas, growing numbers had migrated to Brazil, the Caribbean, and what would become the United States, lured by the warm climates that prevailed there and the thick, rich soils. They built farms, and then plantations, concentrating over time on cotton and tobacco—cash crops that could easily be sold to the elites of Europe. To grow these crops on a commercial scale, they turned to slave labor, importing huge numbers of enslaved Africans through the infamous triangle trade.[12] The impact of this trade was both monumental and tragic, directly killing over one and a half million Africans, and condemning millions more to lives of servitude and misery.[13] In the United States, its long-term ramifications remain, shaping race relations and social discourse in a myriad of ways that no longer have anything to do with cotton. But it was cotton (along with tobacco) that compelled white traders to kidnap or buy millions of Africans and transport them to the plantations of the American South, and cotton that was subsequently shipped—by the boatful, by the ton—back to the ports of England. From there, it catapulted a revolution.

Specifically, cotton created a massive new source of both supply and demand. There was an influx of fibers, obviously—the supply—but also a corresponding demand for skilled weavers who could transform these fibers into cotton cloth. If those weavers could be found, then their production would create another new supply, of cloth, and a corresponding demand for that cloth from a growing population of potential buyers.[14] In eighteenth-century Europe, all the gears on this cycle clicked smoothly into place. Over the preceding decades, the British government had accelerated a process known as enclosure—consolidating small communal land holdings, often by force, into larger private plots. Erstwhile farmers and peasants became as a result a ready supply of labor, moving from the fields into the factories, and tending to looms and spinning machines instead of grain and barley.[15] Agriculture, meanwhile, became more profitable for those who owned the larger farms, and a generation of people who had once spent their lives in musty woolen tunics could now dream of cotton.

And dream they did. By the early decades of the eighteenth century, England was awash in cotton, and Manchester was the center of its trade. English merchants shipped raw cotton to Manchester, then sold it to a rapidly growing network of spinners, weavers, bleachers, and dyers. By 1780, as Sven Beckert notes in his masterful work *Empire of Cotton*, Manchester was the hub of the world's most powerful and profitable industry. "Like Silicon Valley's role as the incubator of the late-twentieth-century computer revolution," he writes, "the idyllic rolling hills around Manchester emerged in the late eighteenth century as the hotbed of the era's cutting-edge industry—cotton textiles. In an area forming an arc of about 35 miles around Manchester, the countryside filled with mills, country towns turned to cities, and tens of thousands of people moved from farms into factories."[16] Many of these factories were homespun affairs, little more than an extra loom or spindle stacked in a cottage's corner.* Many of the workers were women and children, squeezing a few hours of additional labor into a day already filled with chores. But the effect was nevertheless huge, creating both an industrial powerhouse around Manchester and an increasingly specialized labor force.[17]

In 1770, cotton manufacturing had accounted for less than 3 percent of the total British economy. By 1831, some sixty years later, it accounted for 22 percent, and one out of every six workers in the country was part of the cotton trade.[18] In just several decades, as Beckert notes, British merchants had "wrested the empire of cotton from the East within a vigorous generation of invention, rewriting the entire geography of global cotton manufacturing."[19]

What the industry lacked, however, was power. And that's where James Watt comes in.

A Most Singular Invention

James Watt was one of those men who never intended to leave a mark on history. Sickly and shy as a boy, he grew up in the small Scottish town of Greenock, the son of a mathematician who taught navigation and a mother who died young. Rather than being formally educated,

---

* Thus originating the phrase "cottage industry."

Watt went to London in 1755 to study the trade of making mathematical instruments—scales, telescopes, barometers. He returned to Glasgow in 1757 and swiftly set up shop.[20] For several years, Watt labored quietly on quadrants and other astronomical instruments. But then, one day, a professor at nearby Glasgow College asked him to repair a model of the Newcomen steam engine.

As Watt well knew, the Newcomen was a marvel. Developed by Thomas Newcomen, an ironmonger whose personal story has been largely lost to history, it was a relatively roughshod contraption, built around 1700 to solve what had become by that point a most pressing problem: how to pump water from mines. Over the course of the seventeenth century, coal had become an increasingly important source of fuel, particularly in areas like Devon, where Newcomen was born. Denser than wood and more efficient than peat, coal could be burned quickly and easily, firing all kinds of kilns and ovens and furnaces. And there was lots of it in England—not only in Devon but also in Newcastle, and Durham, and along the country's northern edge. Accordingly, coal mines proliferated across Britain between roughly 1550 and 1600, spreading through the countryside and deeper and deeper into its hills.[21] But once mines reached a certain depth, they also almost inevitably tapped into water, flooding the mine and rendering it unusable. To keep the mine going, mine workers had to keep the water out.

But how to do this? Armed with a shovel and bucket, a human laborer could never keep up. The water simply moved too quickly and invidiously, seeping into any cracks or crevices as soon as they were cleared. As early as 1660, therefore, observers had begun to speculate about creating something more powerful, something that could drain the water through more efficient means.[22] They just didn't know how to do it.

What Newcomen had proposed in response was a fairly simple mechanism, based on the realization that water, converted to steam, could be used as a source of energy. The engineering lay in channeling the steam so as to harness its power. Specifically, as the drawing opposite shows, the engine worked by heating water in a closed container, then

Illustration of the Newcomen steam engine, drawn in 1717 by Henry Beighton

drawing the steam that was produced into a separate chamber. As the steam condensed, it created a vacuum, which pulled the piston into the chamber; a series of valves—operated at first by hand and subsequently by a system of strings or rods—then pushed it out again.[23] The up-and-down motion of the piston, propelled by the heating and condensing of the steam, drove a mechanical pump—thus converting steam, for the first time, into power.

By the time Watt began to tinker with his Newcomen model, these engines had proliferated across Great Britain, used primarily as Newcomen had intended them: to pump excess water from coal mines. But the engines were clunky, and expensive to operate. They also didn't lend themselves to operations other than pumping, based as they were on a simple up-and-down motion that required a great deal of coal-fired steam.

Watt's innovation was simple but profound. Working on the model engine, he realized that its basic inefficiency came from the heat that was lost each time the steam cylinder cooled. So he built an external cylinder, which allowed a portion of the steam to cool without chilling the full body of the boiling water. In 1769, Watt filed a patent for his "method of lessening the consumption of steam and fuel in fire engines," and in 1775, he and Matthew Boulton, a far-sighted industrialist, began selling engines. "I sell Here," Boulton promised one early visitor to their works, "what all the world desires to have . . . POWER!"[24]

As had been the case with Newcomen's earlier engine, Watt and Boulton's machine found its first and largest market in the mines. Because the condenser dramatically increased fuel efficiency, mine owners were eager to purchase the new engine, paying Watt and Boulton a royalty equal to one-third of their fuel savings, and guaranteeing the partnership a long and steady source of revenue.[25] In addition, though, Watt's decade of tinkering had enabled him to build a much more versatile engine, one that could be used far outside the world of mining. In particular, Watt's engine could be, and quickly was, tied to England's other burgeoning trade: making cloth out of cotton.

By this point, the scattered looms and workhouses around Manchester had already begun to consolidate, and to approach the early stages of automation. Much of the work was still done as it always had been: by women, mostly, spinning cotton threads on single wheels or weaving the finished threads on looms they powered by foot. But the merchants who purchased the thread and fabric were starting to chafe at the constraints imposed by such a far-flung network of providers, and by the inconsistencies and inefficiencies that inevitably crept in. Quietly, they started to experiment with other forms of production. In 1733, for example, a young mill manager named John Kay invented the flying shuttle—a small piece of wood that weavers could attach to the weft thread (the thread that crosses the warp) and then fling across the loom. The shuttle didn't automate anything, and it still relied entirely upon human labor. But it doubled the productivity of weavers who employed it.[26]

Having more productive weavers, though, didn't necessarily generate greater output, since the weaving side of the trade still relied on the spinners, individual workers each spinning cotton fibers around a single wheel. And so the merchants turned again to innovation—this time in the form of the spinning jenny, a hand-operated wheel that could rotate multiple spindles at once. First it spun eight, then sixteen spindles, tripling the speed of a single spinner and dramatically increasing her production.[27]

The next steps fell into place relatively quickly. The water frame, patented in 1769, used a water wheel situated in a fast-moving river to draw out the cotton threads before spinning. The mule combined the jenny and the frame. And then the power loom, patented in 1783, brought all the new textile technologies into one place, and powered them by steam.

Initially, Watt and Boulton had little interest in adapting their engines for the textile trade. They made pumps, after all, and had built a considerable business and fortune from doing so. By the 1780s, however, with the textile industry spreading rapidly across the English countryside, Boulton had a change of heart. "The people in London, Manchester and Birmingham are *steam mill mad*," he wrote to Watt. "I don't mean to hurry you but I think . . . we should determine to take out a patent for certain methods of producing rotative motion from . . . the fire-engine."[28] Watt resisted, preferring to concentrate his talents on perfecting machines for pumping. In the end, however, he acknowledged that "the devil of rotation is afoot" and set himself to fixing what had emerged as the thorniest design issue of this age: how to use steam power to get a wheel to move.[29]

It wasn't an easy step, either technologically or intuitively. Watt's

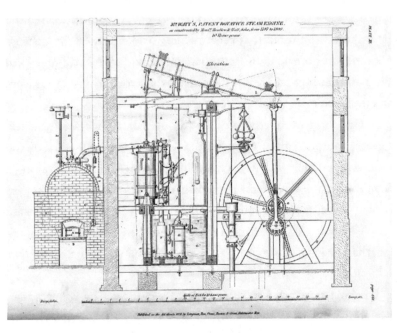

Schematic for Watt's "rotative engine," circa 1800

engine, recall, basically just moved a piston up and down, converting steam heat into a one-way repetitive source of power. Manipulating spindles and looms was a far more delicate task, one that demanded both a consistent source of power and the ability to exercise this power in precise increments. It also meant moving not just a central piston but a series of interlocked gears, propelling them, as Watt later wrote, in a "smoother and more economical cycle."[30] Watt's response to this challenge was what is now called the sun-and-planet, or epicyclic, gear, a configuration in which one central gear drives a series of interlocked gears that revolve around it, much as planets revolve around a central sun. By providing a smooth and continuous motion, this kind of engine could generate the power that the emerging textile industry demanded. And its offspring, as one can see so easily in retrospect, would go on to power steamships and railroads and motorcars and assembly lines— nearly all the machines, in fact, that made the industrial age.

The Rise of the Factory System

By the time of Watt's death, in 1819, Britain's textile industry had become irrevocably and almost completely mechanized. The countryside around Manchester, littered only fifty years earlier with cottage producers and hand-powered looms, was now home to hundreds of full-fledged factories, buildings whirring day and night with workers and machinery and an ever-growing production of textiles. Similar industrial centers had cropped up in France and Germany, some specializing in particular kinds of cloth—silk in Lyon, fine cottons in Alsace—and others simply producing fabric to meet what was now a burgeoning global demand. Cotton, once a rare indulgence for the rich, had become the world's first mass industry, driving a market that, by 1860, accounted for nearly half of all British exports and employed more than 1.3 million people worldwide.[31]

But cotton did more than make a market. It also defined new ways of work and new kinds of workers. Because once textile production moved out of the home and into the factory, once it was driven by steam-powered machines rather than human feet and fingers, the very nature of labor began to change. As Beckert argues, the new machines of cotton "not only accelerated human productivity, but also altered the nature of the production process itself: They began to regulate the pace of human labor."[32]

Specifically, the machines that drove the cotton trade made people move and work faster, at speeds and times that were no longer under their own control. By the middle of the nineteenth century, a typical British cotton mill employed upward of two hundred workers, most of them women and children.[33] Most worked for at least twelve hours a day, six days a week. Some were orphans or residents of the local poorhouses, who worked for free.

What was so transformative about the factory system was not that wages were low or conditions poor, although both of these were true. Instead, what made this system different was that workers, for the first time in history, were tied to both an administrative hierarchy and an industrial one. They worked *for* bosses, in other words, and *with* technology—both of which entailed radical change.

Before the factory system rolled across the north of Britain, most people worked for themselves in at least some sense. Even when they were poor, even when they owed a share of their production to local lords or tax collectors, people eked out a living on their own land and time. In the factory system, by contrast, growing numbers of people worked for someone else—for a boss who owned the factory and profited directly from their labor. As historian David Landes writes:

> The factory was more than just a larger work unit. It was a system of production, resting on a characteristic definition of the functions and responsibilities of the different participants in the productive process. On the one side was the employer, who not only hired the labour and marketed the finished product, but supplied the capital equipment and oversaw its use. On the other side there stood the worker, no longer capable of owning and furnishing the means of production and reduced to the status of a hand.[34]

Today this system is so ubiquitous that it is easy to assume it has existed forever. But it hasn't. The capitalist world in which we operate was forged by the Industrial Revolution, and by the factories it created—factories that brought large groups of people into a single pool of labor and made them accountable to someone else. Watt, presumably, had no intention of reshaping the world so dramatically with his rotative engine, but this is what he did.

Likewise, Watt's engine and the factories that grew around it made workers, again for the first time in history, subject to the whims and demands of their machines. A woman attending to a spinning machine in a Manchester mill, for example, had no ability to slow the speed of her machine, or to shift the way it received the threads she fed into it. She worked for the machine—at its pace, its height, its acceptable range of force and motion. She and her fellow workers had to show up at the factory door at a preset, precise time. They had to work when they were tired and sore, when there were deaths in the family and sick children at home. They had to set accurate clocks and live by them, and find reliable ways of journeying to and from the mills. Once there, discipline was typically harsh: no stopping, no chatting, no singing or whistling above the din.[35]

By the middle of the nineteenth century, the factory system had spread far beyond the bounds of cotton. Automated mines stretched across northern England and along Germany's great Ruhr Valley. Chemical plants in Glasgow and Tyneside were churning out acids and alkalis. Mammoth iron and steel mills were forging the materials that made the machines, and railways linked these networks together and made them run faster still.[36]

To be sure, there were significant differences in how each of these industries was controlled and structured. Some were smaller than others, or more capital intensive, or later in adapting automated techniques. All, though, operated along the lines of the factory system, which, by this point, had become virtually indistinguishable from the industrial age itself.

Divisions of Labor

The Industrial Revolution was a distinct and distinctive moment in time. Before its engines reared across the English countryside, life moved at a slow and steady pace, with the structures of each generation— the ways they lived and loved and worked and married—changing only gradually from what previous generations had known. After the revolution, life changed, jolting several generations into reconfigured patterns of social relations and making change itself ever more of a constant.

Many of the revolution's effects go well beyond the scope of this narrative. As countless accounts have explained, the Industrial Revolution upended business practices and essentially conjured the capitalist system fully into existence. It brought wholesale changes to education, to science, and to the ways in which we approach, and generally embrace, technology. It challenged the dominance of religious authorities and weakened land-based, aristocratic rule. It created global industries, expanded communication networks, and strengthened the foundations for colonial rule. Less obviously, it also changed relations between men and women, recasting the roles they saw for themselves and for the families they forged together.

It was Karl Marx who saw this transformation most profoundly. Writing in the middle of the nineteenth century, and having witnessed the tail end of the Industrial Revolution in Prussia and France, he saw that its effects reached further than the factories and mines that epitomized its spread. "There is one great fact," he wrote in 1856, "characteristic of this our nineteenth century . . . There have started into life industrial and scientific forces, which no epoch of the former human history had ever suspected." As industrial technologies developed, he prophesied, the same forces that were "gifted with the wonderful power of shortening and fructifying human labour" would go on to "starve and overwork it." Sources of wealth would become sources of want, and workingmen—the newfangled creations of the modern age—would rise in revolt.[37]

Marx was wrong, as we now know, about the revolution that would inevitably arise from the turmoil of the industrial age. He was deeply naïve, in retrospect, in presuming that capitalism, once unleashed, could be rechanneled into a socialist utopia in which individuals would cheerfully produce "according to their abilities" and consume only "according to their needs." But he was dead right about the social changes let loose by the factory system.

First and foremost, as he argued, the factory system—and industrial capitalism more generally—created a division of labor that was unprecedented in human history. People had always had different jobs: farmers, artisans, soldiers, or priests. These jobs were somewhat fluid, though, and determined primarily by the class and status into which an individual was born. The factory system, by comparison, took a relatively undifferentiated pool of labor and assigned individuals specific tasks within it: one worker became a carder or spinner, another a weaver. These jobs then defined that worker, and confined him accordingly for the rest of his productive life. Meanwhile, the factory system also embodied the greatest division of the industrial age: that between capital and labor, those who owned the means of production and those condemned to toil for them.

As the industrial age unfolded, Marx argued, a similar division of labor would inevitably settle among human relations as well, reshaping family life along with economic and political conditions. Crucially, men and women would be forced into increasingly separate spheres, with women specializing in home and reproductive tasks and men in more distinctly economic ones. "The first division of labour," he writes, "is that between man and woman for child breeding."[38] Neither Marx nor Engels (who developed this argument more fully over time) viewed this division as evil, or even as imposed consciously by the owners of capital. It was simply the historical result of a specialized labor pool: "when [a woman] fulfils her duties in the private service of her family," Engels explained, "she remains excluded from public production and cannot earn anything; and when she wishes to take part in public industry and earn her living independently, she is not in a position to fulfil her family duties."[39]

The result of this specialization was the growth of an urban proletariat, composed mostly of wage-earning men and their homebound wives. It wasn't a pretty picture. Instead, as Engels wrote in 1884, "this Protestant monogamy leads merely, if we take the average of the best cases, to a wedded life of leaden boredom, which is described as domestic bliss."[40] This was the social order brought forth by the industrial age.

It wasn't just Marx and Engels, of course, who made these predictions, or who observed the profound social change that rocked the eighteenth and nineteenth centuries. Across Europe and the United States—in fact, across all areas affected by the technological change of the Industrial Revolution—society began to shift in fundamental ways, reordering relationships between men and women and reifying their respective roles in the social order. The changes were subtle, and played out differently across the industrial landscape. But their blunt effect was to further the gender divide between men and women, defining ideal types—stereotypes, really—that would quickly and fervently become accepted as the norm.

Die Hausfrau; or, How Technology Made the Housewife

Before the Industrial Revolution, a woman's place was in the home. But so was a man's and a child's—everyone worked close to home, in other words, and chores blended rather seamlessly.[41] Once the revolution hit, however, the workplace was firmly sundered from the home. Cottage trades moved into mills, then factories. Local artisans became urban workers, and the rhythm of the sun and seasons was replaced by the factory bell.

In the initial decades of industrialization, both women and children were an integral part of the expanding factory workforce. Indeed, until around 1830—somewhat later in the United States, and in more recently industrializing countries like Japan and China—women had been the dominant source of labor for the textile trade—young and unmarried women, generally, who streamed into mill towns from Lowell to Lyon and became collectively known as "factory girls." Life in the mills was hard: the hours were long, conditions were cramped, and the work was frequently dangerous.[42] But for some, and particularly in the more liberal-minded mill towns of New England, it was also exhilarating, giving once-cloistered young women a rare glimpse of what independence might be. "I came to Lowell," one former worker recalled, "determined that . . . I would read, think, and write when I could, without restraint."[43]

In retrospect, though, the era of the factory girls didn't last very long. Instead, in country after country, just as the Industrial Revolution was roaring to its peak, men slowly began to replace their wives and daughters, pushing women out of the factories and back toward the home.[44]

Part of what drove this shift was again technology. Ironically, perhaps, as engines became more powerful and automation more complete, factory work became more physical, prizing upper-body strength over manual dexterity.[45] Where an average cotton mill in 1790, for example, would typically have consisted of water-powered spinners, supervised and operated by women and children, its counterpart by 1830 relied instead on heavier, steam-powered machines, operated generally by skilled workmen.[46] Similarly, automation in the mining industry meant a shift away from carrying coal by hand and ladder (a task that had typically been relegated to the wives and daughters of miners) to mechanical pumps and ferries, which were supervised, overwhelmingly, by men.

At the same time, continuous innovation across the agricultural sector also meant that fewer men were needed to perform the chores that had once absorbed their days. Automatic threshing machines, for example, reduced the number of workers needed to separate grains after the harvest; lighter and more nimble iron plows required fewer men to drive them.[47] As this surplus labor drained off the farms, it migrated to the newer industrial sectors. And in an eerie premonition of what would happen more than 150 years later, men began to clamor for what was once "women's work."

Indeed, beginning around 1830, organized groups of men started to fight politically for what they claimed were their jobs. They weren't protesting against women per se, or against child labor. Instead, they were denouncing a system in which factory owners were simultaneously replacing human jobs with machinery and displacing male workers with lower-paid women. In 1835, for example, an organization of English weavers protested against "the unrestricted use (or rather abuse) of improved and continually improved machinery" and "the adaptation of machines . . . to *children*, and *youth*, and *women*, to the exclusion of those who ought to labour—THE MEN."[48] In 1841, arguments before the Short Time

Committee (an influential reform campaign) included pleas for "the gradual withdrawal of all females from the factories."[49] And in the United States, the National Typographical Union similarly urged that "as wages must fall just in proportion to the increase of printers over demand," the hiring of female typesetters would inevitably "depress the liberty and reduce the wages of males." "We do not believe," the (presumably male) Printers' Union wrote, "that any benefit can accrue from taking women from the sphere of action God . . . designed her to occupy."[50]

As these sentiments made their way through the legislative process, both laws and custom began to change accordingly. In the United Kingdom, a series of bills passed after 1844 progressively reduced the number of hours that women and children could work; in the United States, a series of state laws regulating working hours for women would eventually be held up by the Supreme Court. "Her physical structure and a proper discharge of her maternal functions," the court found, ". . . justify legislation to protect her from the greed as well as the passion of man."[51] By 1851, there were practically no "bal maidens"—female mine workers— above the age of thirty-five working in the tin and copper mines of Cornwall,[52] and the female mill workers who remained employed in Great Britain were overwhelmingly young and unmarried. By 1861, fewer than 15 percent of married women in England and Wales remained in the formal labor force; by 1911, that number had fallen to 10 percent.[53]

What accompanied these trends and fed off them was a growing fascination with the nuclear family—mother and father, in a single dwelling, blessed with a handful of healthy children. Again, because this norm has been so ubiquitous until recently, it's hard to imagine that it hasn't been around forever. But this version of the family is in fact a relatively recent creation, born in the later decades of the seventeenth century and cemented by the evolving politics and demography of the late industrial age. Recall, after all, that families in the Middle Ages were more scattered affairs. There were a husband and wife, yes, bound in marriage, but they were liable to be joined at home by unmarried brothers or sisters, widowed mothers or fathers, and children from previous marriages—not to mention the livestock that often shared the family cottage.[54] Because mortality rates were so high and contraception so rare, families also tended to have very large numbers of children, many of whom did not survive.

As the Industrial Revolution unfolded, these patterns shifted in dramatic ways.[55] Mortality rates fell, and dwellings became more spacious. Families could purchase goods—clothing, pottery, cooking implements—that they once had to produce, and could travel across distances that once seemed unfathomable. The spread of public education gave children somewhere to be all day and a universe of possibilities to imagine. For those caught on the wrong edge of the revolution's crush, the effects were devastating: smoke-filled cities, wretched living conditions, and poverty that left people hovering near starvation.[56] It was this suffering that gave rise to capitalism's most eloquent critics, and to the political counterrevolution that would become Marxism.

But, for both better and worse, industrialization had unleashed an economic transformation that could not be stopped. Although scholarship on this topic is both voluminous and contentious, the basic economic facts seem (relatively) clear: in the early years of the Industrial Revolution, the poor of Great Britain likely became poorer, mostly as a result of having been pushed off their lands and deprived of the piecework that had characterized the preindustrial textile trade. After the 1840s, however, both incomes and standards of living began steadily to rise.[57] As E. J. Hobsbawm, one of the preeminent historians of the period (and a lifelong Communist) concludes, "Whatever we may think of the relative position of labourers compared to other classes, and whatever our theory, no serious student denies that the bulk of people in Northwestern Europe were materially better off in 1900 than in 1800."[58]

As incomes thus rose along with the industrial labor force, families began to settle into smaller and more stable units. What we think of now as the traditional family—a married couple, with a breadwinning husband and children who lived at home—became the norm, and then swiftly the ideal across the industrialized world.[59] And what held this ideal together, what defined, in fact, the proper and respectable family, was a nonworking wife. It was her role to mind the children and ensure they received the finest moral education, her role to provide a hardworking husband with a respite from his crowded days and a cozy place of retreat.[60] As the historian Barbara Corrado Pope explains: "Separation of shop and home also meant the separation of men and women and a polarization of familial roles . . . The new factory towns and their squalor, the growing problems of crime and urban poverty, the competition for profit and position, and the possibility of financial ruin characterized the world outside, the man's world." Woman's world, by contrast, quietly exhibited "all those qualities that could not be found in the world outside . . . gentleness and piety, submissiveness and fragility, chasteness and devotion."[61]

Part of what drove this separation was a rise in family income, especially across what was rapidly becoming an urban middle class. Simply put, as men earned greater wages, their wives and daughters could afford to stay home. Even more important, though, was that this economic reality for some members of society became an aspiration for them all.[62] If wealthy women didn't need to work, in other words, and successful men could support their families on a single paycheck, then their family structure became a model for all those who sought to follow in their footsteps. A husband who earned a decent paycheck. A wife who stayed home to support him. And a handful of healthy and productive children, reared to push the next generation of their family to even greater heights. Such became the image of the proper nuclear family, cozy and comfortable and split precisely along gendered lines.

By the latter decades of the nineteenth century, therefore, social pressures conspired with technology to fashion a new sort of feminine ideal: the housewife. She didn't work for wages, this new creature, and didn't venture too far from her home. She was instead wholly a wife and a mother, dedicated to keeping her husband fed, her children well-mannered, and her home immaculately clean. In Great Britain, this ideal was epitomized by the "Victorian angel," a vague but popular image of the good woman—pious, submissive, and devoted to her domestic chores, "immensely sympathetic," in Virginia Woolf's words, "immensely charming [and] utterly unselfish."[63] In the United States, it was the True Woman—devout, again, and obedient, and working as hard on the home front as her husband did on the job.[64] And in Germany, it was *die Hausfrau*, whose thrift and cleanliness (often referred to as *solide Häuslichkeit*, or "solid domesticity") ensured her family's place among the nation's bourgeoisie.[65]

These images varied, as one might expect, along with the cultures that had produced them. But they shared a deeply similar vein of aspiration and advice, showing women and girls what they should want to be, and how to get there. The core of the image was moral piety—virtue, as it was often called, wrapped up with submissiveness and domesticity.[66] The good wife, the Angel, was sweetly religious, and devoted to her husband and children. She tended to their needs, fed their desires, and bestowed her purity upon them. Instead of sullying herself with the hustle and bustle of the outside world, she presided over home and hearth, doing whatever was required to keep them gleaming and well stocked and warm. As John Ruskin, a prominent British social theorist, wrote in 1865, "The man's work for his own home is to secure its maintenance, progress and defence; the woman's is to secure its order, comfort and loveliness."[67] The job of the Angel—and it was distinctly a job—was to "set out accurately, and in something like scientific order, the laws which govern, and the rules which should regulate, that most necessary and most important of all human institutions, THE HOUSEHOLD."[68]

There was a central contradiction in all these images, of course, since the woman who was responsible for these household chores was rarely depicted as actually dirtying her hands with them. But the image was just that: the ideal, the illusion, the fantasy and aspiration. Many women of the industrial age exhausted themselves trying to fulfill the demands of house and home. Many hired servants to do their dirty work, and many women without other means went into service themselves, cleaning and polishing households that others could claim as their own.[69] Yet the image—of the good wife, the housewife, the Angel—remained, defining what women should aspire to, regardless of their actual state in life.

One of the most telling artifacts from this era is a series of "household manuals" launched roughly around 1860. The central one, and the icon, was Isabella Mary Beeton's *Book of Household Management*, a thousand-page tome that reportedly sold sixty thousand copies in its first year of publication.[70] According to Beeton, "Of all those acquirements, which more particularly belong to the feminine character, there are none which take a higher rank, in our estimation, than such as enters into a knowledge of household duties; for on these are perpetually dependent the happiness, comfort, and well-being of a family."[71] Similar sentiments were expressed in French (*Guide des femmes de ménage, des cuisinières, et des bonnes d'enfants*) and in German (*Schweizer Frauenheim*), the latter focusing in particular on the pride that emanated from a tidy and well-stocked linen closet.[72]

Seen through the prism of time, such advice sounds quaint.* But it revealed a considerably deeper truth. Housework *was* work, and for families without flocks of domestic servants it was an arduous, complicated endeavor, made more arduous and complicated by the social pressure that had been thrust upon it. During the Middle Ages and right through the time of the Renaissance, housework was invisible, something that happened but was rarely commented upon. Now it had become a sign of social status, and a tangible measure of a woman's worth. As historian Deborah Simonton notes, "In many respects the family home became for a wife what business became for her husband. A housewife's job was to maintain and direct a well-run household, in the same way as her husband ran a shop or business."[73]

Both men and women were thus working what were essentially full-time jobs. Both were increasingly connected to a global capitalist economy, and to the technological innovation that drove and defined it. They were working differently, though, perhaps more differently than

---

* Although perhaps not that exotic, given that similar advice is still regularly dispensed.

had ever before been true. A cascade of technologies had fallen upon them—steam, engines, railroads, and all that was enabled in their wake—and social structures had changed, as they always do, in response.

To be sure, the connection between steam power and housewifery is neither simple nor direct. Watt's engine didn't conjure housewives into being, any more than plows created the adoration of virginity. Instead, revolutionary changes in technology brought about equally revolutionary changes in how societies produced their goods and structured themselves. The plow led slowly but inexorably to agriculture and private property, the steam engine to factories and industrial capitalism. As these changes rippled and surged through society, they drove corresponding changes in family life—changes that simultaneously supported the new modes of production and legitimized them.

Let me underscore the causal point here. My argument is not that particular social structures were necessarily preordained by the technologies of the Industrial Revolution. On the contrary, one can imagine factories that might have stayed female throughout this transformation, or countries that embraced a greater gender balance as they industrialized. (Indeed, and not coincidentally, both the Soviet Union and China could be described this way.) The point is that family structures and gender roles don't exist in or emerge from a vacuum. Instead they come from a complex interplay of ostensibly unrelated factors, most of which are sparked or at least accelerated by technological change. We make the machines, once again, and then they make us.

Engines of Change

The Industrial Revolution never really came to an end. Instead, its technologies continued to steam and simmer around the world, spreading, eventually, to all but the farthest-flung regions and changing in the process every culture they touched. By the time World War I exploded in 1914, nearly all of Europe had entered the industrial age, along with Japan, China, India, and huge swaths of Asia and Latin America. Factories were simply a way of life by that point, while agriculture—which had dominated and defined human existence for the past eight thousand years—was being pushed increasingly toward the social and economic sidelines.

If we compare this revolution to its agricultural precursor, the biggest difference, perhaps, is speed. The technological change that occurred in the Neolithic valleys of the Tigris and Euphrates crept slowly around the planet, moving from settlement to settlement across thousands of years. The inventions of the eighteenth century, by contrast, moved in decades, occasionally even in years. Watt created his steam engine in 1769; by the early 1830s, over one hundred thousand power looms had been installed, lumbering away not only in Great Britain but in France and Germany and the United States as well.[74] British manufacturers built the first automated cotton mills in 1784, but the technology slipped swiftly—sold, copied, or stolen—across international borders, creating within decades mill towns and factory systems that completely upended whatever had existed before. Capitalism as we know it came, too, providing the financial system (and, many would argue, the accompanying political system) that allowed the wheels of industry to keep turning.

All of these changes came quickly, unfolding in less than one hundred years. If you plot human civilization on the proverbial twenty-four-hour clock, with the Agricultural Revolution commencing at 12:00 a.m. and today as midnight twenty-four hours later, then the Industrial Revolution struck at 11:35 p.m., only minutes before the iPhone in the broader sweep of history.[75] In retrospect, it struck so quickly that people living through it lost perspective on the past. Did the advent of steam—of engines and factories—create a world for eighteenth-century workers that was better than what their parents had known, or worse? Was it the herald of progress or decline?[76] Was it, to echo Dickens, the best of times or the worst of times? Marx's synthesis or antithesis?

The answer, probably, is both. All. Everything at once. Because the revolution turned everything upside down and topsy-turvy, frequently in ways that weren't entirely linear. Automation ripped peasants from the fields and thrust them into factories, and then threw them out again when the next round of automation reduced the need for labor. Industrial capitalism challenged the power of land-based aristocracy even as it created new and perhaps even more impregnable structures of control. Women thronged to join the paid labor force even as they were told that virtue lay at home. Men ruled their households and lost control of their lives at work. It was an era of vast wealth creation, and of grinding, desperate poverty. The best of times, in some ways. And the worst. No wonder, then, that Marx foresaw that the very same forces that pushed women home during the Victorian era would eventually push them out again, armed over time with the power to protest.[77]

In retrospect, what changed most during the Industrial Revolution was the notion of change itself. Before the revolution, people's days played out mostly as their parents' had, with livelihoods and lifestyles determined by slow-moving technologies and even more stagnant social structures. Afterward, people presumed—without even being explicitly aware of their presumptions—that their lives and opportunities would be different from those of their parents, and that their children's lives would be more different still. A girl born in, say, 1700 outside of Manchester would have lived a fairly predictable life. She would have grown up in a small cottage, milked cows and spun woolen yarn by the fire, and walked pretty much everywhere she ever traveled. She would have died, typically, around 1740, never having ventured more than five or ten miles outside her village, or having seen anything that would have been new to her parents, or their parents before them.[78] That same girl born in 1800, by contrast, would have seen factories and smokestacks. She might have worked in a cotton mill, and experienced both freedoms and burdens unimaginable to her parents. She would have witnessed steam-powered machines, boats, and locomotives—all new, all unprecedented. She would still probably have married, and borne children, and died before the age of forty. But she would have grown up experiencing change and expecting it to continue.

This was the impact—the seismic psychic impact—that the Industrial Revolution had wrought. Before men made the machines of the eighteenth and nineteenth centuries, technology had evolved so slowly that life stayed stable and predictable. Afterward, change itself became a constant, whipsawing communities from one generation to the next, and imposing expectations we now take as norms. Growth. Progress. A division of labor between classes and people. And a sense—fainter in some times and places, stronger in others—that individuals can shape the futures they face.

BY THE TURN of the twentieth century, the revolution had rolled out of Manchester, and steam itself yielded to the higher powers of gas and electricity. But Watt had captured a process that far outlived his engine: Make a machine. Catalyze an industrial change. And watch it pulsate throughout our lives.

3.

Mid-Century Modern

Like the breeze in its flight, or the passage of light,
Or swift as the fall of a star.
She comes and she goes in a nimbus of dust
A goddess enthroned on a car.
The maid of the motor, behold her erect
With muscles as steady as steel.
Her hand on the lever and always in front
The girl in the automobile.

—Frank Leslie's Illustrated Weekly, 1907[1]

The American kitchen of 1962 was a beautiful thing to behold. High gloss and newly efficient, it boasted a range of gadgets and appliances that looked like mini-missiles and heralded a future of household bliss. There was the Creda Carefree electric cooker, which promised to switch itself on and do the cooking for you. The Westinghouse Heavy Duty Laundromat. And the Hoover Constellation vacuum cleaner. All available in avocado green and harvest yellow, hues as hip as the technologies behind them. "So much of your life is spent in the kitchen!" cooed one appliance ad from the time. "It's simply good sense to key your kitchen to modern living . . . Food will taste better and you'll have so much more time—and pep!"

Kitchens at this point in time, along with laundry rooms and bathrooms and other once-quiet corners of family life, had become spectacles of public consumption, signs that a family had achieved a certain level of prosperity and sophistication. They were also technological marvels, gleaming with gadgets that would have been unimaginable only a few decades earlier. Even at the height of the Industrial Revolution, women typically spent long days and weeks laundering clothing and preparing meals. Now piles of dirty towels could be tossed into the Hoovermatic twin-tub washing machine, doused with a capful of Rinso washing powder, and left to spin merrily on their own. Dinners could be purchased in neat little containers—beef with gravy, hashed brown potatoes, a side of corn, and cherry crisp!—popped into the Wedgewood Electra-Matic double oven, and consumed by the happy family clustered around the Zenith television.

The line between Watt's steam-powered engine and the Hoovermatic washer is not completely obvious. Indeed, we tend to think of industrial technologies as remaining and replicating entirely in that realm, with the steam engine leading to the combustion engine, for example, and the engine to the railroad. All of which is true. Yet industrial technologies also tend to trickle into the more intimate realm of consumer goods, creating and redefining our most basic tools and desires. At the turn of the twentieth century, this trickle became a flood, letting loose a torrent of consumer items—cars and washing machines and toasters and contraceptive pills—that utterly transformed daily life.

To be sure, many other technologies were invented and adopted during this busy moment in time. Nuclear weapons were discovered and deployed; radio and television reshaped the landscape of mass communications; electricity spread power around the globe. But this chapter focuses on a trio of technologies whose impact on gender, sex, and the family were particularly profound. Automobiles gave women and young people the newfound freedom to move. Household appliances liberated wives, at least in theory, from the drudgery of unpaid work. And the birth control pill gave women autonomy over their sexual and reproductive lives. Together, cars, appliances, and contraception changed how individuals worked, lived, and loved.

Riding Around in Your Automobile:
The Revolution That Was the Model T

The most obvious change came from the automobile, perhaps the most pervasive symbol of mid-century modernism. In the span of less than three decades, cars brought the muscle of the Industrial Revolution directly into the lives of thousands and then millions of drivers around the world. It was cars, in fact, that turned people into drivers, giving them physical, tangible control over forty horsepower of steel and rubber and gas. Linger on that word "driver" for a moment to realize its import. Before the car, people moved—and conceptualized movement—in terms of the horses they had been riding and pulled by for centuries. They moved, literally, at the horse's speed and under its power. Cars took the power of these beasts, multiplied it severalfold, and put that power in the hands of eager men and women, giving them an unprecedented jolt of both mobility and freedom. Cars put people in the driver's seat and gave them places to go.

In their earliest incarnation, passenger cars had been around since the late nineteenth century, when a number of tinkerers and technical entrepreneurs had independently seized upon the idea of attaching an internal combustion engine to the formerly horse-drawn carriage.[2] These early models were expensive and clunky, however, and exceedingly hard to drive; they were playthings and gadgets for a while, but hardly a realistic substitute for horses, carriages, and trains. Then, in 1902, an engineer named Henry Ford started to build cars for the masses. Technologically, Ford's car was more or less like those of the other innovators in the field.

His genius, though, was to realize that by reducing the costs of production he could also lower the price and expand demand for what were still luxury goods.[3] "I will build a car for the multitudes," Ford prophesied early in his career, "constructed of the best materials, by the best men to be hired, after the simplest designs that modern engineering can devise."[4] Through automation, Ford realized, he could create a mass market for cars.

Accordingly, just after the turn of the century, the Ford Motor Company introduced the Model T, a car specifically engineered to be purchased and driven by ordinary people. It was cheap—$850, or $950 with a windshield and top—light, and easy to operate.[5] You simply turned a key in the ignition slot, cranked the engine into operation, and then maneuvered the vehicle through a combination of three spring-loaded floor pedals and a steering wheel.[6] Almost immediately, the car was a hit, selling rapidly among farmers, small tradesmen, and moderately salaried employees.[7] As demand grew, the company's Highland Park Plant in Detroit sprang into action, adding more assembly lines and labor to its highly automated manufacturing process. Between 1910 and 1916, Highland Park grew its output from 32,000 cars to 735,000—a stunning feat for a product that had only recently been considered a luxury item for the leisure class.[8] Holding true to Ford's prediction, Highland Park then used this surge in demand to reduce its costs of production, and the sales price for the Model T. By 1916, a basic Model T—and they were all basic: black and simple, with identical four-cylinder, twenty-horsepower engines—cost only $360, or the equivalent of about $8,500 in 2019.[9]

Initially, the market for these machines was presumed to be almost entirely male. Cars were tools, after all, smoke-belching, crank-turning tools that lumbered outside the home and helped men ply their trades. Cars were dangerous to drive and difficult to master. They were meant, most turn-of-the-century observers insisted, for men. "The natural training of woman," one 1909 article claimed, "is not in the direction to allow her properly to manipulate an automobile in case of emergencies."[10] "Women," the author continued, "do not very commonly possess the nervous imperturbability which is essential to good driving . . . They are too easily worried, too uncertain of their own right of way, too apt to let their emotions affect their manipulation of the steering wheel."[11] Not surprisingly, many commentators also argued that female drivers posed a threat to the existing social order. On the road and behind the wheel, women drivers were seizing expertise that was meant to be men's, and stealing precious time from their own, more homebound tasks.

Yet, as car ownership surged across the United States, it turned out that women were indeed learning to drive. Quickly, and in droves. In fact, within just years of the Model T's introduction, women were getting behind the wheel in large numbers, learning to navigate a new and faster-moving terrain, and rapidly embracing the autonomy that came as a result. One farm woman, for example, whose story was described in the *Rural New Yorker* of 1919, began to leave the evening meal to boil in a "fireless cooker," so that she could wash the breakfast dishes, drive to visit her daughter, and run a butter and egg route along the way. Another began to carry her farm's milk to the creamery each day in the family car, using the time she saved to indulge in a few pleasure rides each week and even an occasional trip to the movies.[12] Another, asked why her household had purchased an automobile when they couldn't afford a bathtub, was brisk and precise: "Why, you can't go to town in a bathtub," she replied.[13]

By the late 1930s, women in the United States were driving as frequently as men. And as their presence became ever more commonplace, the link between driving and autonomy—between steering a motor vehicle and seizing real personal power—was similarly made and reinforced. "There is a wonderful difference between sitting calmly by while another is driving," wrote one early enthusiast, "and actually handling a car herself. There is a feeling of power, of exhilaration, and fascination that nothing else gives in equal measure."[14] "Whenever I feel nervous or out of sorts," confessed another, "I get into my car and drive off my troubles."[15] The Ford Motor Company, ever prescient in these early days of the revolution it had wrought, captured this power shift in its marketing materials. "It's a woman's day," one advertisement promised. "She shares the responsibilities—and demands the opportunities and pleasure of the new order. No longer a 'shut in,' she reaches for an even wider sphere of action—that she may be more the woman. And in this happy change the automobile is playing no small part."[16]

As this "happy change" played out across wider swaths of the American middle and working classes, it began to wreak massive change in the country's urban geography as well. Within decades, the spread of cars across once-rural areas began to transform many of these areas into suburbs—new communities perched between the cities and the countryside, and linked to both of them by the ever more ubiquitous auto and an accompanying network of paved roads and highways.[17] By 1939, private cars accounted for nearly 90 percent of all travel between American cities, multiplying the range of places that people could visit, commute to, and live.[18] By the end of World War II, as returning soldiers hastened to the suburbs to build the lives they had put on hold, this new geography morphed into new patterns of social aspiration as well, carving a distinctive lifestyle that quickly became the American Dream. It was green and clean, this suburban utopia, shaped by single-family houses along shaded streets and populated by hardworking men, doting wives, and 2.5 towheaded kids. It was innocent and idealistic, upwardly mobile and implausibly pure. And it was powered, always and inevitably, by cars. By 1960, 15 percent of American families owned two cars or more, and the U.S. auto industry, based largely in Detroit, accounted for nearly 4 percent of the total American economy.[19]

As has been recounted elsewhere, the impact of the automobile industry was vast. Both in the United States and around the world, cars created jobs and wealth; they built industrial centers and pumped pollutants into the atmosphere. Quietly, though, and less obviously, they also continued to reshape the contours of family life.

Most directly, for those women with the means and ability to drive, cars quickly became both a burden and an escape, simultaneously another thing they had to do and a means, in theory at least, of escaping it all. Because once families had moved away from the city, someone— and it was usually the stay-at-home mother—had to drive.[20] Someone had to ferry the children to school or football practice, drop visitors off at the train station, and haul groceries from the market. In some respects, therefore, the car defined a new stay-at-home role for suburban women, thrusting them into nearly full-time jobs as the family chauffeur. But because women tended to have control over their families' cars, they also gained independence and mobility. After the chores were done, women could use their cars to take them wherever they wanted to go. Cars were private, after all, and they didn't tell tales. With cars at their disposal, housewives could begin to imagine driving their own destiny.[21]

Meanwhile, the spread of cars in America and elsewhere also shifted the balance of power between young people and old, and particularly between parents and the teenagers they had once aspired to control. Before the car, there really were no teenagers. Young people just matured gradually into adults, studying or working under the watchful eye of their parents until they married and moved fully into families of their own. The Industrial Revolution started to change these patterns, creating new opportunities for people to earn wages away from home, and new means of transportation to get them there. But it was cars that jolted this change into something far more powerful and widespread; cars that gave young people both a symbol of their independence and a physical space in which they could both experiment and escape.[22]

During the nineteenth century, across both the United States and Europe, unmarried adults had very few opportunities for physical intimacy, much less sex.[23] Young men could go to brothels, or risk a passing kiss (or more) with a barmaid or a household servant. Young women—at least of the respectable sort—had virtually no such options. Instead, they were expected to interact with men only in public, and in ways their elders deemed socially acceptable. A young man wanting to "call" on a girl had to do so at her home, with her parents or other chaperone in constant attendance.[24] Grabbing a kiss under these circumstances was fairly tough; anything more could prove socially disastrous.

Beginning in the 1920s, though, young men started to call on young women in cars and then drive away unattended—to the speakeasies that flourished in the United States during Prohibition, to diners that were beginning to populate rural roadways, to secluded lovers' lanes that could be reached only by driving.[25] It may be too much to trace the suddenly relaxed social norms of the 1920s—the flappers and jazz and short skirts and hair bobs—just to the arrival of the automobile, but the causal connection is clearly there. As one sociologist notes somewhat dryly, "The opportunity for sexual expression was made possible in the anonymity of . . . the automobile."[26] Before long, gaining a driver's license had become a rite of passage, a tangible sign that a young man or woman had crossed the Rubicon into adulthood. And gaining access to dates and kissing and the possibility of sex had become an implicit part of the bargain.[27]

Constructing Convenience; or, How the Refrigerator Changed the World

Inside the home, meanwhile, another explosion was occurring: a blast let loose by the end of World War II, powered initially by the rising prosperity of the American family, and dispersed, within decades, to include most of the developed world.

This revolution began more subtly. Unlike the case with cars, where Ford's technology burst quickly and raucously across the national landscape, household appliances began their march to modernity more quietly, and generally in the unsung shadows of the kitchen, bathroom, and laundry room. Unlike cars, household appliances sat firmly and unequivocally within the realm of women, and never experienced anything like the popular excitement that surrounded the Model T. In their own way, though, refrigerators—and washing machines, vacuum cleaners, and toasters—were every bit as revolutionary as automobiles. They liberated household workers from the drudgery of their tasks, giving them time at last to tend to other things. And because the vast majority of these workers were women, it was the refrigerator, truly, that set them free.

Ever since the Industrial Revolution, women had been spending the bulk of their lives tending to family and home, cooking and cleaning and mending most everything that existed within their house or cottage. These were massively time-intensive tasks. At the turn of the twentieth century, even a comfortably middle-class woman could reasonably expect to spend roughly eight hours a day on basic household chores: going to market; sewing and mending clothes; heating water for baths and cleaning; pickling, canning, and preserving food. For all but the wealthiest wives, therefore, being a workingwoman meant being a housewife. There simply was no time left for anything else.

The first technology to make a dent in this workload was the icebox, an early-stage refrigerator that simply housed ice, which made its commercial debut in the 1870s. Prior to this point, keeping food fresh had been a perennial, time-sucking process. During the summer and fall, women spent hours gathering perishable produce and canning it for the colder months. They boiled fruit into jams and jellies and dug cellars in which to store potatoes and other root vegetables. They smoked meats, butchered chickens, and either gathered produce or ventured to the local market nearly every day. With an icebox, though, these chores were much diminished. So long as the iceman delivered his haul, women could purchase perishables like milk, meat, and butter and store them, for days or even weeks, right in their own kitchens.

But the icebox was limited, of course, by its reliance on ice—a bulky, short-lived commodity that needed to be replenished every week or so. So it was a very big deal in 1916 when the Packard Motor Company began to produce the Isko, a "wonder machine" designed to "fit any ice box" and keep it chilled, without a need for ice.[28] The Isko didn't work very well, and sold only fifteen hundred models, but its underlying technology was sound, using the evaporation of a contained liquid to cool the area around it. As a generation of innovators (including, briefly, even Albert Einstein) tinkered with the mechanical and chemical processes involved, home refrigeration became smoother and more reliable, maintaining a constant temperature inside the once far-more-unreliable icebox, and even manufacturing the luxury of ice. Reporting on this novelty, even the usually unflappable *New Yorker* was stunned: "A little water is put in some mysterious place [inside the refrigerator]," the magazine reported, "a few minutes pass, a magic door opens, and a tray of small ice cubes appears before startled eyes."[29]

The next great advance was the washing machine. In 1907, the Hurley Machine Company introduced a basic home machine called the Thor. It had the laundry room's essential wood tub, yet connected now to something radical—an electrical motor, powered by the grid that was rapidly expanding across the United States and elsewhere, and capable of operating a series of pulleys and gears to shake and tumble whatever clothes were deposited in the tub.[30] The Thor had no way to heat the water inside the tub, and its external motor created the constant risk of shock, yet still it found a market, a small but eager group of buyers happy to be relieved of at least the most arduous aspects of laundering. Over the next few decades, these basic innovations grew sleeker and more profound. Thor's external motor was moved inside (by other firms). Water heaters were added, along with more sophisticated means of tumbling and even "damp drying" the clothes. In 1937, the Bendix Home Appliances company introduced a full-scale automatic washer, complete with washing, draining, rinsing, and spinning cycles. With the Bendix, housewives could simply deposit their dirty laundry into the machine, set its automatic controls, and, as an extensive print advertisement promised, enjoy their "new leisure, freedom from toil, new economy, and far cleaner clothes."[31]

Looking back from the relative ease of the twenty-first century, it's hard to imagine just how revolutionary this was. Because today we still complain about doing the laundry. We still lug smelly clothes around in bulky bins, and find old socks abandoned on the stairways. So it's useful to recall what doing the laundry entailed before the days of the automatic washer. For starters: a housewife needed to make lye to help remove stains or soda to keep her washing water soft, manufacture or procure starch, and maintain more than fifteen washing-related articles. She had to sort and soak all the clothes the night before washing day, and then re-soak, boil, and rinse them all by hand, before wringing them out and hanging them to dry.[32] Even with additional help—a frequent practice, given the sheer physicality that laundering involved—washing day was exhausting, absorbing hours and hours of a housewife's time, week after week after week. The Bendix and other machines of its ilk changed this equation in a fundamental way.[33] Because even though its capacity was tiny by contemporary metrics (nine pounds, compared with up to twenty-eight pounds for top-loading washers today), even though most housewives still relied on outdoor lines to dry their laundry, and even though standards of cleanliness remained looser than they are today, au-tomatic washers revolutionized the task that was laundry, saving their owners hours of sweaty, heavy work each week. No wonder, then, that sales of appliances soared in the immediate postwar period, generating record profits for the firms that made household conveniences and estab-lishing them as some of the world's most iconic brands: Westinghouse, Whirlpool, General Electric, Maytag, Electrolux.

There is some academic debate over exactly how much time these manufactured marvels saved the average housewife. Not all families could afford washing machines or vacuum cleaners, of course, and as historian Ruth Schwartz Cowan has argued, some middle-class women may initially have done more work than their mothers or grandmothers, since the rush of appliances replaced the servants or commercial services that once handled the heaviest tasks.[34] But the overall data clearly de-scribe a march of appliances into American households and a reduction of hours thus devoted to household labor. By the early 1940s, 79 percent of households in the United States had electric irons, 52 percent had wash-ing machines, roughly half had refrigerators, and 47 percent had vacuum cleaners.[35] Spaces that were once sites of hard labor—the kitchen, the laundry, the bath—had become instead showpieces of modern efficiency, crowded with time-saving appliances and whirring with the constant flow of electricity. Nearly all American homes were hooked to the country's expanding power grid by 1950, and women were spending far less time on what were once their daily tasks. In 1900, an average housewife spent nearly sixty hours each week just tending to the basic tasks of her home. By 1975, that number had plummeted to twenty-three, freeing up nearly a full workweek formerly devoted to labor for other, outside pursuits.[36]

Just how women would use these extra hours, though, was not entirely clear. As Cowan and others have noted, the early years of the appliance revolution were marked by a certain cultural disconnect, a quiet tension caused by the fact that machines like the refrigerator and washing machine both reduced women's work and made it more attractive. With thirty-five newfound hours each week, a once-housebound wife could theoretically leave the drudgeries of laundering and cooking and pickling to forge a career outside the home. Or she could simply revel in an easier home, one made cleaner, brighter, and more perfect by the tools now at her disposal. Household technologies, in other words, drew women into the seduction of their kitchens even as they freed them from its rigors. And in the process, they gave women a subtle but transformative choice: whether to embrace their inner housewife or discard her.

Certainly, as feminists from Betty Friedan to Naomi Wolf have pointed out, the companies that sold these toasters and dryers and freezers and irons were eager to convince mid-century women that their appropriate place was still at home.[37] Indeed, the focus of their advertising—and advertising budgets soared during this period—was to convince women that housework had become somehow sexy, the source of feminine pride rather than the simple drudgery of yore. Print ads from this era purr with a certain well-oiled luxury, intimating a not-so-subtle link between a woman's appliances and her marital happiness. One 1962 ad for a "sheer look" washing machine, for example, announces that "a good washer is like a good man."[38] Another proclaims that "you should choose a refrigerator the way you'd choose a husband."[39] The route to the bedroom, it seems, runs straight through the kitchen.

Advertisement for Bissell sweepers from *Better Homes & Gardens*

Economically, this made great sense. Because appealing to women's lust for tools—selling fantasies of perfectly cleaned cooktops and sweet-smelling socks—created a vast consumer market in place of what was once cheap labor. Between 1947 and 1971, the market for "white goods" (as economists came to label appliances) grew from $60 billion to over $160 billion, and a rash of new products emerged as well, promising to make the appliances' results even shinier, fluffier, and more sweet smelling.[40] NuSOFT Fabric Softener for the washing machine. Cascade for the dishwasher. And Pop-Tarts for the toaster. These were the staples of family shopping, and the icons of American consumerism. And in order for sales of these items to remain high, women had to be convinced to keep buying them, and to indulge in the specific labors they embodied. They had to love their new machines and see their own identities—as women, wives, and mothers—mirrored in the gleaming appliances that now graced their homes.

By the 1950s, therefore, kitchens in America had become both items of aspiration and temples of feminine artistry. Wealthy women—even middle-class ones—could spend their days not only tending to their families but indulging them as well, with picture-perfect pot roasts or ornately molded salads. They could have "what women want" (Tide detergent, apparently) or prepare meals that "look better, taste better, are better," feeding, perhaps, their own creative whims as well.[41] Working-class women, meanwhile, could fantasize about ending their drudgeries in ways that were at least imaginable, even if not easily attained. In *Little Shop of Horrors*, for instance, a fabulously dark musical about the era, the poor shopgirl Audrey dreams wistfully of a life away from Skid Row. She wants love, yes, and children along the way, but her description of perfection is an appliance seller's dream: she yearns for a suburban tract house with a "disposal in the sink / A washer and a dryer and an ironing machine." And why not? After centuries of hard labor and few rewards, technology had made housework fun and creative and maybe even a little bit sexy. Armed with the right kind of tools, even poor girls like Audrey could dream of a life—or at least a lifestyle—that was once reserved solely for the rich.

Yet not all women wanted to devote their extra hours to either the fanciest of tools or the cleanest of kitchens. Some of them wanted instead to flee those kitchens and, in the words of Betty Friedan, "see housework for what it is—not a career, but something that must be done as quickly and efficiently as possible."[42] For these women, and their daughters and granddaughters to come, the appliance revolution was more than a change in lifestyle. It was the chance to change their lives.

By the early 1960s, and despite a global culture that still doted on the merrily mopping stay-at-home mom, more and more women had quietly reentered the workforce.[43] As Cowan and other authors have noted, most of these women didn't actively challenge the societal norms that still surrounded them or fight, as they might have, for other, more communal ways of distributing household labor.[44] They didn't disrupt the middle-class vision of a breadwinner husband married to a homemaker wife. The vast majority of them earned far less than their male colleagues, and held far more menial jobs. But quietly, and then seismically, women were surging into the workforce. By 1970, 37 percent of the U.S. labor force was female, and 30 percent of married women with children under the age of six worked outside the home.[45] Forty percent of Australian women had entered the workforce by that same year, as had 50 percent of Japanese.[46]

These changing patterns of women's work are well-known and are generally attributed to the waves of feminism that swept across the United States and much of Europe in the 1960s and '70s. Yet feminism, like most social movements, was shaped and enabled by technology. It was pushed into prominence not only by the fiery rhetoric of its founders but also by technological advances that, for the first time in centuries, made it entirely possible (if not yet easy) for women to work both inside the home and away. As technology freed them from labor, in other words, so, too, did it transform women's place in the household, and in the world. And thus the lowly Maytag not only washed the towels. It became a force for liberation.

Better Living Through Chemistry: The Power of the Birth Control Pill

The third set of technologies that paved the way for feminism were perhaps the most profound. Because if cars gave women the means to escape their homes, and appliances gave them the time to do so, it was birth control that freed them at last from their most enduring form of labor: conceiving, birthing, and caring for babies.

The prehistory of birth control is as obvious as it is bleak. Ever since humans started to cluster around the communal fire pit, after all, women had been constrained, far more than men, by their reproductive biology. Women were the ones who bore the children and nursed them, the ones who conceived at the hands of men and the will of Mother Nature. With the exception of a few scattered and relatively inefficient means of birth control—abstinence, the rhythm method, and condoms made from sheep's intestines—they had virtually no means of preventing pregnancies they did not want or babies who came at unfortunate times. All the major shifts in family structure, including the move to monogamous marriage and the rise of the nuclear family, had only narrowed women's place in the world, focusing as they did on the sanctity of family life and the division of labor between men and women.

Birth control shook this state of affairs to its very core. Between 1957 and 1973, an estrogen-based contraceptive known simply as "the pill" spread from the poorly funded laboratory in which it was hatched to become the world's single bestselling pharmaceutical product. The pill was a technical marvel—safe, effective, and massively profitable. It was, and remains, one of the most popular and frequently prescribed of all modern medications, the precursor to "wonder drugs" like Lipitor and Viagra. Its greatest impact, though, was both deeply personal and massively political. Because the pill changed women's lives forever.

What the pill did most radically was separate sex from reproduction, pleasure from pregnancy. For the first time in history, women could deploy technology independently and inconspicuously to have sex as they pleased, without necessarily committing themselves to either marriage or children. And this technology—slowly but inevitably, linked as it was to cars and washing machines and the other tools of mid-century modernism—began by the end of the twentieth century to reconfigure all that prior generations had taken for granted about families and gender and sex. It wasn't feminism, in retrospect, or changing social mores that gave rise to the pill. On the contrary: it was the pill that launched the revolution.

Bitter Herbs and Potions:
Birth Control Before the Modern Era

Although the historical record is spotty, the search for contraception appears to be nearly as old as conception itself. The methods varied from place to place, as did the circumstances under which women were urged or permitted to use them. In ancient Egypt, for example, women were advised to use various pessaries and tampons, concocted from mixtures of acacia, colocynth, and dates. Ancient Hebrews preferred lower-tech methods, such as coitus interruptus supplemented by "magical potions" or "performing violent abdominal movements after coitus." Neither was particularly effective.

As Christianity gradually gained sway over the Western world, even these basic methods were generally condemned, subject to religious and eventually legal norms that defined conception as sacred and contraception therefore a crime.[47] Population records suggest that birth control of some sort was used at least sporadically in Europe during the Middle Ages, since most women bore fewer children than sheer biology might suggest.[48] But, in general, for women across space and time and geography, babies came of their own accord, regardless of their mothers' needs or wants. And women, in turn, were doubly bound: to the men whose babies they conceived, and to the children they then bore.

In the nineteenth century, as population levels surged across the Western world, a handful of enlightened thinkers began to ruminate publicly on the prospects for population control. In 1822, for example, a man named Francis Place—a working-class father of fifteen—published "Illustrations and Proofs of the Principle of Population," one of the first modern documents to explicitly advocate birth control. Several years later, a doctor in Massachusetts published *Fruits of Philosophy*, the first book by a physician to deal explicitly and encouragingly with birth control. Looking, he later recalled, for "some sure, cheap, convenient, and harmless method, which should not in any way interfere with enjoyment," Dr. Charles Knowlton urged young married couples to use a syringe filled with water and any salt or astringent to kill the "small and delicate animalcules" and "destroy the fecundating property of the semen."[49] Copies of the book sold wildly, and Knowlton soon built a prosperous medical practice. But, like Place before him, he had only really succeeded in publicizing the problem of contraception. He hadn't provided a way of solving it.

What came closer, finally, was technology—here in the form of vulcanized rubber, the creation of an inventor named Charles Goodyear who had no particular interest in birth control. Instead, Goodyear was simply trying to combine natural rubber, a flexible but volatile substance, with something that would prevent the rubber from cracking in cold temperatures or melting in warm ones. After he succeeded in 1839, rubber became a booming business and companies such as B. F. Goodrich sold hundreds of rubber items, including ropes, tires, boots, and a suitcase that could double as a life preserver. Quietly, and without fanfare, they also sold condoms, listing them discreetly in their catalogs as "male shields" or "rubber goods for gents."[50]

For women, however, the subject of birth control remained almost completely taboo. Nice girls didn't talk about it, didn't condone it, and certainly didn't admit to practicing it. Medical doctors frequently blamed contraception for causing all sorts of illnesses, and self-styled "purity crusaders" emphatically linked contraception to prostitution, condemning both as unnatural acts committed against the family.

For women trying to moderate the size of their families, however, the biggest obstacle was practicality. Yes, condoms helped. And rhythm, practiced diligently, sometimes worked. But neither of these methods was anything close to foolproof, and neither put decision-making power in women's hands, or bodies. And so, well into the twentieth century, women were having sex more or less as their ancestors had—guarding their virginity until marriage, staying faithful afterward, and tending to whatever babies Mother Nature sent their way.

This state of affairs prevailed until 1960, when Searle launched the pill that would change the world.

The Business and Science of Birth Control

In scientific terms, the pill's development really began in the 1920s, when researchers first identified two powerful female hormones, progesterone and estrogen, and began to track their joint role in conception and pregnancy. For decades, scientific research into their function was dominated primarily by those seeking to understand, and eventually cure, infertility.[51] But as this once-murky science began to be better understood, a handful of researchers began to realize that the process could also be engineered in reverse. In other words, if the right combination of estrogen and progesterone could coax an infertile woman's body into conceiving, then an inverse combination could potentially, and more easily, prevent an otherwise fertile woman from conceiving. Because contraception still lingered in a legal and moral limbo, however, researchers had little interest in pushing their discoveries toward a more applied end.

Matters didn't change much until 1952, when Katharine McCormick, a wealthy and well-connected widow, decided to enter the game. In 1904, McCormick graduated with a degree in biology from MIT, the second woman to have done so. She then married Stanley McCormick, heir to the International Harvester Company, who suffered, doctors soon discovered, from a severe case of schizophrenia. During the subsequent decades, Mrs. McCormick used her wealth and scientific background to fund neuropsychiatric research that she hoped would help her husband. She also tried, repeatedly, to avoid his sexual demands, determined not to pass his schizophrenia on to her offspring. When Stanley died, in 1947, McCormick switched her philanthropic focus and directed all her funds toward the pursuit of an oral, "female" contraceptive. At the urging of

Margaret Sanger, a passionate and longtime advocate for medical contraception, she put her money behind a controversial geneticist named Gregory Goodwin Pincus, whose work on parthenogenetic, or "test tube," rabbits had brought him both scorn and notoriety in the 1930s. Denied tenure at Harvard, Pincus established the Worcester Foundation for Experimental Biology in 1944, and settled down to work on other, less provocative projects. His first love, however, was still researching the mechanisms of reproduction, and when Sanger and McCormick approached him with funding, Pincus jumped at the opportunity to work on an oral contraceptive.

Between 1952 and her death in 1967, McCormick gave Pincus and his foundation nearly $2 million in research funding, a princely sum for a single, risky project. And by 1955, it had already paid off. What Pincus was able to do, remarkably quickly, as it turned out, was to demonstrate that a simple daily dose of progesterone combined with estrogen could essentially fool a woman's body into believing that she was in the early stages of pregnancy. Accordingly, women who took this dose of hormones—in a formulation quickly known as the pill—could prevent themselves from conceiving a child. They could have sex with whomever they desired, whenever they desired, without having to fear becoming pregnant. Moreover, because the pill was tiny and discreet, because it left no discernible effect on a woman's body or menstrual cycle, no one had to know if a woman was "on" the pill, or even if she was sexually active. Almost as soon as Pincus had proven the efficacy of his work, a small pharmaceutical company named G. D. Searle began to plan for its commercial introduction, slipping stories of this medical breakthrough into publications like *Ladies' Home Journal* and *Vogue*, and quietly touting the pill's effectiveness to a growing circle of doctors.[52]

In 1957, the U.S. Food and Drug Administration approved Searle's application to manufacture Enovid, the country's first commercial birth control pill, as a treatment for "gynecological disorders." By 1961, an estimated 408,000 American women were on the pill; by 1963, the number had risen to 2.3 million. Similar surges occurred around the world, with millions of women embracing chemistry now as a means of regulating fertility.[53]

In the process, Searle also converted contraception from a social taboo into a cultural icon. With the advent of the pill, birth control became an enabling technology rather than a furtive hope, the symbol of women's freedom from their physiology. Once the technology of oral contraception found its way to the market, women across the world were liberated in wholly unprecedented ways. Liberated from their reproductive biology. From social stigma. From their family and clergy and fears of both abortion and unwanted pregnancies.

On a deeply personal level, the birth control pill changed not only women's individual lives but also patterns of behavior more generally. As the economists Claudia Goldin and Lawrence Katz have noted, the pill altered the marriage market in the United States by enabling young men and women to delay marriage without having to forgo sex.[54] Marriage rates thus fell in the aftermath of the contraceptive revolution, while the incidence of both premarital and extramarital sex rose. Similarly, because women no longer faced such a stark trade-off between early-marriage-with-children-and-sex and career-without-sex-or-children, more women were willing to invest in their careers, confident that they could still have sex without risk and get married, if they chose, at a later date.[55] As a result, women entered the professional workforce in greater numbers and stayed there longer; they surged into law, medical, and business schools.[56] Men's behavior changed, too, for both better and worse. On the one hand, as Goldin and Katz note, it made some men more willing to "wait" for women who invested in their own careers and education; on the other, as Katz and his colleagues George Akerlof and Janet Yellen find in a separate paper, by lowering the risk of premarital pregnancy, it removed the long-standing societal expectation that men would marry the women they "got in trouble." In the pill's wake, they write, "Men who wanted sexual activity but did not want to promise marriage in a case of pregnancy were neither expected nor required to do so."[57]

In many ways, therefore, the birth control pill upended the entire edifice of family controls that had existed since the time of the plow. Women didn't need to bear the stigma and burden of unwanted pregnancies. Men didn't need to fear the marital obligation. And thus, in terms of pure demographics, the advent of contraception began to do what many conservative critics had feared: to slowly chip away at the contours of the traditional family structure.[58]

MEANWHILE, SOMETHING ELSE was happening in the wake of the contraceptive revolution, something more subtle but equally important in the context of this book. To put it most simply: the nature of technological change itself began to evolve. It was faster, for one thing. More immediate. And it hit sex, gender, and family life far more directly.

Recall that during each of the last great revolutions, technological innovation begat social change through a clear but derivative path. Breakthrough inventions like the plow and the steam engine changed the economic infrastructure of the societies in which they developed—Marx's means of production—and then these economic changes catalyzed a corresponding set of social transformations. So the plow gives rise to private property, which shifts the economic value of women and children and reshapes the norms of marriage. The technologies of the Industrial Revolution enable the factory system and industrial capitalism, which in turn force a redefinition of men, women, and families. At each of these transition points, a seismic shift in technology calls forth an equally seismic shift in social structures, but the mechanisms of change are gradual, and run through the complex medium of the economy.

With contraception, however, we pass an inflection point, since this is the first truly radical technological innovation to affect the individual before it affected the state, the market, or any greater community. The pill's revolution begins, quite tangibly, in the human body. It affects life even before the moment of conception, changing not the means of *production*, but instead and even more radically, the means of *reproduction*.

And thus the causal mechanisms swerve. Instead of following the path of the plow and the steam engine, whereby changes in sex, family, and gender are shaped and filtered first by changes in the economy and industrial infrastructure, the pill forged changes from within. With the pill, in other words, as with genetic engineering and assisted reproduction and other innovations yet to come, the seeds of change may lie, quite literally, within us.

Manufacturing the Modern

On a typical day, I don't think much about technology. If I do, it's usually because something (the remote control, most frequently) isn't working, and I need to find an able-bodied family member to help me out. When I do think about technology in my personal life, I think about "technology"—that is, about inventions that have not yet come to pass, and that raise intriguing questions I can't quite yet answer. Would I want my body to be cryogenically preserved? Maybe. Give up my trusty Subaru for a self-driving car? Perhaps. Like most people, in other words, I think of technology as what lies ahead, rather than what shaped the reality I already inhabit.

But, because I am a professional woman in the early decades of the twenty-first century, a wife and mother and daughter and friend, my entire life has been shaped and enabled by technologies that didn't exist a century ago. Without a car, I could never have commuted daily to my first job as an assistant professor. I certainly couldn't have managed to get three kids to three different schools on my way to the office, or grab groceries from the market on my way home. Without a dishwasher, a washing machine, a vacuum cleaner, and an oven, I couldn't have balanced my family life with my professional one. Instead, like generations before me—and along with the millions of women without the means to purchase some household conveniences—I would have been forced to choose between my family and my job, between taking care of my children and earning the income to buy both them and myself some greater measure of independence. And without access to the pill and other modern forms of contraception, I wouldn't have been able either to limit my family to a manageable size or to time my children's births to a fairly manageable moment. I am therefore a product of technology in everything I do, in everything I am. So is every human ever conceived, born as they are—as we all are—at a particular moment in history, and carved and shaped by the innovators who created our past and paved the way for our futures.

The new machines of the mid-twentieth century seem almost quaint today, hardly even worthy of an innovator's label. But cars and refrigerators and contraception are what made our last century modern, and what created the kinds of social patterns that we now think of as normal: working mothers, suburban families, and sex that often has nothing to do with either marriage or procreation. None of these now-commonplace things is either natural or preordained. In fact, they're not even particularly old. Instead, they are the social remnants of a single century of technological change. And as we embark on another century, one marked by ever-accelerating rates of innovation, these once-modern structures are bound to evolve yet again.

Part II

The

Way

We

Live

Now

4

Changing the Means of (Re)production

Surrogate Moms, Gay Dads, and Our Evolving Notions of Family

If reproduction is once completely separated from sexual love, mankind will be free in an altogether new sense.

—J. B. S. Haldane, *Daedalus, or, Science and the Future*, 1923

The body . . . is the source of our problem. If we didn't have wombs, we'd be fine. It's about reproduction.

—Gloria Steinem, 2017[1]

Some months ago, in the space of one week, I received two very similar emails. One was from a former research assistant, the other from a faculty member. Both are lovely men, but not people to whom I am particularly close. Both were writing with happy news—that they, along with their new husbands, were in the process of becoming dads. One had already completed what he termed the "process" and was looking forward to his daughter's impending birth. The other was still negotiating the purchase of eggs and considering, as he noted with some trepidation, the most recent offer from the fertility clinic: "Jewish, runway model, Ivy League grad, for only $100,000!" Surrogate not included.

This is the new face of the not-so-nuclear family. It's older, it's frequently led by either a single parent or a gay couple, and its children have been hatched through nonsexual means. Mothers, in many cases, have become pregnant through a combination of purchased sperm and in vitro fertilization; fathers are employing surrogate carriers with whom they have a financial rather than sexual relationship. It's a whole new world.

Throughout most of human history, the link between sex and children, marriage and sex was so tightly strung as to be only inconceivably separated. As described in this book's first section, marriage itself was a way to control women's fertility and ensure paternity: if women were virgins before marriage, and remained monogamous thereafter, then children came naturally (most of the time), families formed predictably, and life proceeded as it should.

The advent of the pill first shattered this cozy structure in the late 1970s, making way for both feminism and women's march into the workplace. But contraception—preventing pregnancy—is only half the equation. The other half—*producing* pregnancy—is more recent and even more radical. And it is exploding at a dazzling pace.

Since 1978, when a British housewife named Lesley Brown gave birth to Louise, the world's first "test-tube baby," the field of assisted reproduction has been unpacking the technical means of procreation, increasingly replacing sex with science. Today, twenty-five-year-old women entering law school (or, quietly but more commonly, thirty-five-year-old women mourning the end of a bad romance) can, for around fourteen thousand dollars, plus three hundred to twelve hundred dollars a year in storage fees, freeze their eggs in perpetuity, hoping to thaw and use them at some later, better time. Forty-five-year-old women whose fertility has long since diminished can purchase the frozen eggs of younger women and use them to become pregnant. Lesbian couples can purchase sperm online; gay men can buy eggs, hire surrogate carriers, and—for around one hundred thousand dollars—become fathers to their own genetic children. Increasingly, would-be parents can use technologies such as intra-cytoplasmic sperm injection (ICSI) and pre-implantation genetic testing to cherry-pick their children's genetic characteristics and protect against possible mutations. The growth of these technologies has generated considerable hand-wringing over the future of our children, and a fear that we are on the cusp of creating a superrace for the superrich. But the bigger issue—and a probability that has already slipped comfortably into reality—is that we are reshaping the fundamental nature of what it means to breed.

Bringing children into the world was once a task reserved for a distinct, albeit rather large, slice of the human population. Babies came to young, married, heterosexual couples. They were cared for, overwhelmingly, by the young, married, female half of this couple. This relationship no longer holds true. Instead, babies can now be produced by and for a much wider array of people—single women, single men, gays and lesbians, women well beyond their natural boundaries of fertility, and even, in extreme cases, through the genes of parents no longer alive. There are issues of commodification in all this—we have been buying babies since the days of Louise Brown. There are issues of objectification—when we buy, we do indeed select for tall and pretty and smart. The biggest issue, though (and a surprisingly underexamined one), is how assisted reproduction is bound to transform our fundamental notion of the family. Because once we no longer need the family structure—one man, one woman, one set of ovaries and some sperm—to create children, the family as we know it is almost certainly destined to evolve. Not for everyone, of course. Many couples will hew happily to traditional norms of heterosexual marriage and babies conceived through sex. But increasing numbers of people—paired and single, gay and straight—will form their families through science instead. And this crucial shift in the means of reproduction will herald a corresponding shift in how we think of and define the so-called normal family.

In earlier technological revolutions, the mechanisms of intimate change ran through the economy—innovation engendered social structures like private property and the factory system that in turn reshaped the contours of daily life. With assisted reproduction, by contrast, as with contraception before it, the effect is much more direct. By changing the way in which even a small fraction of people conceive their children, reproductive technologies will inevitably shatter some of our most fundamental notions of sex, love, and family. Indeed, they already have.

The Science of Sex

In most areas of innovation, it is the technologies that are sexy. Think of the steam engine, thrusting its way across the mines and mills of the eighteenth-century landscape. Or the streamlined, turbocharged machines of the mid-twentieth century. Observers reveled in these inventions' power and prowess; users bathed in that certain glamour that comes from being smart and forward thinking and bold.

By contrast, nobody really likes to talk about the science of assisted reproduction. It feels unnatural, somehow, a violation of both the sacred and the profane. Because almost everyone can produce babies. Everyone always has—easily, and usually enjoyably. Inserting technology into humankind's oldest and most basic equation seems cold. Awkward. And decidedly unsexy.

But for nearly a hundred years, technology has nevertheless been making its way across the reproductive landscape. As of 2017, roughly four hundred thousand children across the world were being conceived each year through some form of assisted reproduction—artificial insemination, egg donation, surrogate wombs, in vitro fertilization, and pre-implantation genetic testing.[2] In some parts of the world, where the technology is particularly advanced and governments subsidize its use, the percentage of artificially conceived children is above 3 percent, and growing rapidly.[3] Look around any playground, schoolyard, or toddler's birthday party. One of those kids—more, in affluent places—was conceived through nonsexual means, with petri dishes and pipettes playing the intimate role that was once reserved for human bodies.

The science of nonsexual reproduction began slowly in the nineteenth century, when a handful of doctors began to question the source of their patients' infertility. Infertility itself was nothing new. Indeed, across time and space and race and place, some significant portion of all human beings—roughly 10 to 15 percent—have been unable to conceive children through normal sexual means.[4] Until recently, however, cases of infertility were treated simply as a bad luck (or, in some cultures, as signs of wickedness in the women so affected). "Barren" women were subjects of sympathy or scorn; they could be divorced by husbands seeking to produce heirs, or seek solace in the care of other people's babies. They couldn't do anything, though, to address the underlying cause of their infertility.

Around the middle of the nineteenth century, however, some curious physicians in the United States and England began to suspect that their infertile female patients might not be the sole cause of their own infertility. These doctors were treating young, recently married, apparently healthy women. They were also, in some cases, treating their husbands, who had suffered from premarital bouts of syphilis. Until this point, infertility was always presumed to be the woman's "fault": men, medical reasoning had long held, could not be infertile unless they were physically incapable of ejaculation.* But these doctors started to wonder whether their male patients' venereal diseases could perhaps be the source of the problem. To test this theory, they engaged in one of modern medicine's most outrageous experiments: secretly inseminating their female patients with the sperm of other men.

The best-known of these procedures occurred in Philadelphia, where, in 1884, a physician named William Pancoast offered to treat a young woman's infertility with a newly developed operation. She agreed, he anesthetized her, and then he slipped a vial of sperm from

---

* Which, as it turns out, isn't even true. Impotent men are not necessarily infertile.

one of his medical students into her womb. The woman subsequently gave birth, apparently never knowing either the genetic origins of her child or the fact that an early bout of gonorrhea had rendered her husband—and not her—infertile.[5] The story came to light only in 1909, when one of Pancoast's former students published a short article in *The Medical World*, describing what had happened on the operating table.[6]

By this point, though, others had begun quietly implementing what would soon become more widely known as artificial insemination, using sperm from fertile donors to impregnate women who could not conceive with their husbands. It was a covert procedure for decades, restricted to married heterosexual couples and usually kept a secret from the children they produced. But artificial insemination worked, and by the 1970s commercial sperm banks had started to emerge across the United States, Western Europe, Australia, and Israel, offering a wide range of possible donors. For around twenty to thirty-five dollars a vial, patients could choose the genetic material that would become their child: blond hair or brown, blue eyes or green. One bank even infamously went so far as to offer only the sperm of Nobel Prize winners and other official geniuses, tempting recipients with visions of producing a little Einstein or Newton.[7] By 1980, roughly twenty thousand babies a year were being conceived through artificial insemination.[8]

The next great breakthrough in reproductive science—and still the most important—came in 1978, when Lesley Brown gave birth to the five-pound, twelve-ounce "test-tube baby" Louise.

Technically, Louise's conception had nothing to do with a test tube. The baby sprang, instead, from a procedure known as in vitro fertilization (IVF), which had been pioneered several years earlier in the experimental laboratories of Robert Edwards and Patrick Steptoe, researchers at England's Centre for Human Reproduction. For nearly a decade, Edwards and Steptoe had been tinkering with different combinations of egg and sperm, different methods of mixing the microscopic packages of cells and coaxing them into the dance of reproduction. They experimented with a range of fertility-enhancing drugs and endured, along with their patients, a long stretch of miscarriages and ectopic pregnancies. And somehow, after more than eighty failed IVF procedures, it worked. For the first time in history, an embryo created outside a human body was born, forty weeks later, as a wholly formed child.

The news of Louise's birth was greeted with a particular mixture of awe and horror. Outside observers raced to condemn "test-tube babies" as inherently unnatural—as an immoral, abnormal intervention of humankind into nature's realm. At the University of Chicago, for example, bioethicist Leon Kass argued publicly that "this blind assertion of will against our nature—in contradiction of the meaning of the human generation it seeks to control—can only lead to self-degradation and dehumanization." Paul Ramsey, a leading Protestant ethicist, similarly pronounced, "Men ought not to play God before they learn to be men, and after they have learned to be men they will not play God."[9] Even more damaging criticism came from the Catholic Church, which was unequivocal in its opposition to any form of assisted reproduction. "From a moral point of view," the Vatican proclaimed in 1987, "procreation is deprived of its proper perfection when it is not desired as the fruit of the conjugal act, that is to say of the specific act of the spouses' union."[10]

In the end, though, it didn't really matter. Over the next twenty years, as the technologies of IVF became more and more reliable, millions of would-be parents rushed to embrace them. By the spring of 1983, only five years after Louise's birth, roughly 150 babies in the United States had been conceived in vitro. That number soared to 8,741 by 1993, and to nearly 50,000 by 2003.[11] By 2006, more than 6.5 million babies around the world had been conceived with the help of IVF.[12] Presumably, none of the parents involved had undergone religious transformations or changed their minds about sex. Instead, they chose to substitute science for nature simply because nature wasn't working for them, and science was. And once science had worked its magic—once those cells in the laboratories of antiseptic fertility clinics had morphed and multiplied into embryos and fetuses and perfectly normal children—the science itself began to feel ever more natural.

By the 1990s, a handful of IVF clinics had begun to experiment with another process that once seemed impossible—replacing one woman's eggs with another's. Unlike sperm, eggs are difficult to reach. They cannot be delivered outside the body, and even the most fertile woman in the world produces only one or two each month. But as clinicians began to explore the possibilities for egg transfer—quickly dubbed "egg donation"—they realized that the problems were not particularly intractable and the possibilities vast. Essentially, by using the technologies of IVF, they could simply harvest eggs from one woman, fertilize them in the laboratory, and then transfer them to the womb of another. Once these steps worked, and they generally did, a woman could give birth to a child who was not genetically hers, a child who otherwise could never have been conceived. Or as one early recipient reported, "I have just given birth to a miracle child."[13]

By the early 2000s, egg donation had become a key component of assisted reproduction. Women dealing with infertility—and particularly those over the age of thirty-five—were regularly advised to turn to donor eggs after only a few failed cycles of IVF, and younger women were being urged to freeze their eggs early, in case they either wanted to use them later in life or donate them to someone else.[14] Egg donor agencies had proliferated across the Internet and in those geographical regions where IVF use was particularly robust and commercial. Much of this activity was directed at what could by this point be called traditional IVF—married heterosexual couples, combining donated eggs with the husband's sperm to conceive a pregnancy that the wife would carry, resulting in a child they'd call their own. But increasingly, egg donation had also enabled other, more radical, permutations. Most crucially, it was being used to transform and expand the market for surrogacy.

Until this point, science hadn't created many options for women who couldn't carry pregnancies or men who didn't have the physical means to do so. Occasionally, couples who were desperate to conceive would quietly turn to surrogates—women willing to be inseminated with the husband's sperm and then carry a child to term. But such cases were rare and fraught with difficulties. Because the surrogate was also the genetic mother of the child she bore, women occasionally changed their mind after giving birth, plunging the couple and their surrogate into pitched and emotional battles.[15]

Egg donation changed this equation in fundamental ways. Now, using donor eggs, a couple could purchase eggs from one woman and procure a womb from another. Gay couples or single men could hire a surrogate to give birth to a child conceived with another woman's eggs. By combining donor eggs with what became known as gestational surrogacy, would-be parents had a far greater range of procreative options than ever before, and a broader array of technologies from which to choose.

Meanwhile, another band of researchers was developing what would come to be known as pre-implantation genetic diagnosis (PGD).* Unlike other means of assisted reproduction, PGD wasn't created to assist with reproduction. It was, instead, a highly technical, hugely experimental means of addressing genetic diseases like Tay-Sachs or cystic fibrosis—horrible, usually fatal diseases caused by a known mutation of a single gene. As the genetic foundations of these diseases became better understood, specialists started to imagine testing for them before a child was even born, creating some way for couples who knew they carried genetic risk to produce and select embryos that didn't carry the disease.

In 1989, two researchers in London—Robert Winston and Alan Handyside—figured out how to do the science. Trying at that point simply to identify the sex of an embryo (and thus eliminate certain sex-linked diseases such as hemophilia), they managed to remove a single cell from a set of three-day-old embryos, identifying each one as

---

* PGD is also increasingly referred to as pre-implantation genetic testing (PGT).

either male or female. The following year, they repeated the procedure, this time selecting female embryos that were subsequently transferred and brought to term as healthy baby girls.[16] Working with Winston and Handyside, an American biologist named Mark Hughes took the next step: analyzing early-stage embryonic cells for a wider range of genetic mutations. In 1990, Hughes and Handyside used this pre-implantation diagnosis to help a couple who carried the gene for cystic fibrosis give birth to a healthy child. And in 1999, Hughes worked with a transplant surgeon in Minnesota to create an even more dramatic conception—Adam Nash, born after several rounds of IVF and PGD, and selected from dozens of embryos on the basis of a specific genetic marker. His umbilical blood was taken at birth and used to save his dying sister.[17]

As had been the case with IVF, news of PGD raised specters of superbabies and genetically modified children. Writing in 2002, the bioethicist Leon Kass argued that "the price to be paid for producing optimum or even genetically sound babies will be the transfer of procreation from the home to the laboratory . . . Such an arrangement will be profoundly dehumanizing."[18] His concern was echoed by political scientist Francis Fukuyama, who worried that "when the [genetic] lottery is replaced by choice, we open up a new avenue along which human beings can compete, one that threatens to increase the disparity between the top and bottom of the social hierarchy."[19] What actually happened in the labs and clinics, though, was decidedly more mundane. Between 1990 and 2000, over a thousand would-be parents used PGD to create embryos that were specifically modified to eliminate genetic mutations such as Tay-Sachs or Huntington's disease.[20] Several thousand more used the technology in the course of ordinary IVF treatments, asking technicians to select the healthiest-seeming embryos, or those that passed most robustly through a screen of genetic tests.[21] Ironically, though, as the tests became increasingly sophisticated, concern over the procedure died down. By the turn of the twenty-first century, analyzing your child's genes before they entered the womb had become almost commonplace. Like artificial insemination before it, like frozen eggs and test-tube babies, PGD was just another way to have a baby, using technology to supplant what had once been left to nature.

Family Affairs

Like many breakthrough technologies, assisted reproduction was conceived without revolutionary intent. William Pancoast wasn't trying to change the nature of parenting when he inseminated his apparently infertile patient; Edwards and Steptoe had no desire to uproot the family structure when they perfected the process of IVF. On the contrary, all three were working in a distinctly conservative format, using technology to help parents produce what is perhaps the most nonrevolutionary product of all: a child to call their own.

As these technologies evolved, however, they also began to be used in nontraditional arrangements, and by people seeking to create very different forms of family. By using assisted reproduction, for example, a lesbian couple could conceive and parent a child without having to involve a third-party dad. A single woman could become a mom without the entanglement of a man. And, most dramatically perhaps, a gay male couple could combine a donor's eggs with a surrogate's womb to give birth to a child of their own. In each of these newfangled combinations, the child would still be the genetic product of only one of its intended parents. But for the first time in history, that child would not need to be the product of a heterosexual union. Instead, men could create babies without having sex with women, and women could create them without having sex with men. And thus what began as a wholly conservative desire—to help married, infertile couples produce the children that nature denied them—became the radical foundation of something else. Of new forms of parenthood, and families, and sex.

To get a glimpse of how these changes are playing out, I met in 2017 with some of the families at the technology's forefront, families whose lives are already a hodgepodge of the truly revolutionary and completely mundane.[22]

Michael Hammer's dining room, for instance, doesn't look like a launchpad for change. There is a big table that serves as the center of family life, with a kitchen island surrounded by books. Several cookbooks are lying open and dog-eared on the island, and the family's impressive collection of Islay whiskey bottles beckons from the shelves above. Michael's twins, Zoe and Josh, are in the adjoining living room, doing homework on their laptops and watching YouTube videos. The usual. Yet Zoe and Josh aren't quite ordinary twins. In fact, as Michael describes it, they're more like "half twins . . . born on the same day at the same time from the same uterus" but with different genetic fathers and a rather unconventional conception story.

In 2002, Michael and his now husband, Eran, began the arduous process of trying to become dads. Determined to produce kids who would be genetically related to them, the two men found a surrogate in California willing to carry their child, and a separate egg donor. They went through two rounds of IVF—mixing both their sperm each time with the donor's eggs—and holding their breaths through what would eventually be three failed pregnancies. Then they switched to a new donor and a new surrogate, did a "fresh round of everything," and finally welcomed the twins in 2004. Today, the two dads don't know or care which of the kids is related to whom. "My children are probably only half related," Michael explains. "Traditionally, you had two types of twins, identical and fraternal, but now there is a third type, you can have half a twin, and that's perfectly natural."[23]

AJ, my former research assistant, and his husband, Adam, are earlier in the process. A couple now for fifteen years, they have carefully outlined what they want in a surrogate—a "reasonable person," AJ says, "with a good head on her shoulders and a support system"—and in an egg donor. They have lined up their doctors and a therapist, a donor concierge (to help with the selection), and a number of different surrogacy agencies. The two men are looking forward to their eventual children, and to sharing the stories of their birth. "I mean, nobody in this family has a uterus," AJ says with a laugh, "so it is what it is."[24]

As AJ is quick to acknowledge, however, the very fact that he and Adam can contemplate parenthood so blithely is a recent and radical development. When AJ came of age in the late 1980s, gay men were only beginning to come out of the closet. Gay marriage was unheard of, and gay parenting difficult to imagine. Yet technology has now pushed the once-impossible nearly to the edge of ordinary. And in the process, people like AJ and Michael, simply by pursuing that most basic of desires—a child to call their own—have begun to reshape the structure of society.

Legally Yours: Assisted Reproduction and the Changing Laws of Parenthood

Traditionally, we haven't thought of parenthood as a legal situation. It's a messy operation, a sleep-deprived state, a massive commitment of love and time and energy. But, until recently, becoming a parent didn't seem to involve any legal intervention at all. Two people simply turned to nature when the time was right and produced an offspring who instantly became theirs. Doctors might be present at the moment of birth, but lawyers rarely were.

Assisted reproduction has changed this state of affairs in a massive way, not only by bringing lawyers into what was once a deeply private process but also, and more fundamentally, by challenging the underlying construct of parenthood. For centuries, nearly all civilizations have deemed a child's parents to be the married mother and father who conceived him. She who physically bears the child is his mother, tradition tells us. He who is married to the birthing woman is his father. These conventions run strong and deep across the world's major religions and legal codes—so strong and so deep, in fact, as to have attracted virtually no scrutiny over time. Instead, we have accepted and enshrined into law the basic geometry of the married heterosexual couple; in other words, because the only way to produce a child was through heterosexual sex, we took that natural process and made it law as well.

As we have started to conceive children through other means, however, the old laws have begun to fray. Because what if two women want to raise a child together but only one gives birth? What if the birth mother is a surrogate who has no intent to parent the child? Traditional law has no way to address these situations, other than to prohibit them. Which is exactly what most laws did when the technologies of assisted reproduction first appeared. But as these technologies have become increasingly successful, the prohibitions have broken down. And in their stead is slowly emerging a brave new world of what it means to be a parent, and a child.

Bastards and Other Daddy Issues

To trace the evolution of these laws, it's useful to begin with something that has nothing to do with technology and everything to do with sex.

Ever since time immemorial, women have given birth to children who weren't conceived with their husbands, children who were instead the product of rape or romance or extramarital lust. And ever since humanity invented the plow and embraced legal marriage, the children of these unions have been treated differently.[25] As bastards. Illegitimate. An economic burden and social shame. In sixteenth-century England, illegitimate children could not inherit property or take their father's name, and were often either killed or abandoned; in Sweden, an illegitimate child was considered a foundling, with no legal mother or father.[26] There was a certain irony to these prohibitions, since most women—or at least most married women—could easily ascribe their children's conception to their legitimate husbands. There also seems to have been a dramatic disregard of them, since geneticists estimate that somewhere between 5 and 15 percent of all men listed on birth certificates are not the genetic fathers of their children.[27] But laws against illegitimacy prevailed for centuries, bolstering a broader societal embrace of monogamy and only-marital sex. Even when women had sex outside marriage—which they did—and even when they produced children from these liaisons—which they did as well—the law enforced and upheld the dictates of legitimacy. Legitimate children were conceived only inside marriage, the product of one man and one woman, linked by a legal document and bound by something like love. Everyone else was a bastard.

Matters began to change, though, around the 1950s, as rates of illegitimacy surged across the United States and Europe. Perhaps it was the spread of automobiles and the freedom they provided; perhaps it was that women had more time on their hands and fewer fears about the stigma of unwed motherhood. In any case, the rise was real and well-documented. Between 1938 and 1958, the estimated number of illegitimate births in the United States increased dramatically, pushing the illegitimacy ratio (the estimated number of illegitimate births per one thousand live births) from 38.4 to 49.6.[28] In England and Wales, that same ratio rose from 54 in 1941 to 85 in 1968.[29]

As these numbers pushed upward, they both generated and coincided with a newfound concern for single mothers and their children. This shift in attitude does not have any technological antecedent. Instead it came from the general awakening of social consciousness that would roil the 1960s, from a new generation of reformers who sought to protect unwed mothers (who were often poor and of color) and to shield their children from the long-term effects of generational poverty. In terms of this book's argument, therefore, the fight to redefine illegitimacy does not quite fit. It is about ideological shifts rather than technological ones, and about family formations that did not arise from any discernible changes in the modes of production or reproduction. But because this struggle carved important precedents for redefining the "legitimate" child, and because these precedents would subsequently be applied to define and protect the children of assisted reproduction as well, I am including it here.

The story begins in 1960, when the U.S. state of Louisiana implemented a "suitable home" law, withholding public assistance from unwed mothers on the grounds that they were morally incapable of providing appropriate care.[30] The law itself was fully in line with traditional practices in Louisiana and, under normal circumstances, might simply have settled within the contours of that state's social policies. But these were not normal times in Louisiana, or across the American South. Instead, and against the backdrop of the civil rights movement, it suddenly seemed unfair—at least to a group of energetic reformers—to treat fatherless children as a separate and less privileged class. Accordingly, as soon as the law was passed, a young lawyer named Melvin Wulf wrote a brief to the U.S. Department of Health, Education, and Welfare, stating, "We believe that any differential treatment to which out-of-wedlock children are subjected is invidious and likely unconstitutional." Referring to *Brown v. Board of Education*, the landmark 1954 Supreme Court decision that denied states the ability to treat people differently on account of their race, Wulf argued that "differential treatment accorded a class of citizens designated as 'illegitimate' is equally baseless."[31] Instead, Wulf emphasized that the sins of the parents, such as they were, should not be allowed to affect the well-being of the child. "It is clearly unreasonable," he urged, "to attempt to improve a home's moral climate by starving its occupants to death."[32]

Meanwhile, in Illinois, a law professor named Harry Krause was also pushing the case against illegitimacy, arguing that a child born out of wedlock had no possible way of affecting the circumstances of his or her conception. Why, then, should the child suffer for the parents' sins? Why should anyone be subject to what he labeled "the psychological effect of the stigma of bastardy"?[33] It was time, he insisted that "the matter be considered from the standpoint of the child!"[34]

And, very shortly, it was. In 1968, the U.S. Supreme Court considered the question of bastardy in *Levy v. Louisiana*, a case involving the alleged wrongful death of Louise Levy, an African American mother of five who died after a Louisiana hospital failed to treat her symptoms of what turned out to be kidney failure. Initially, Louisiana courts dismissed the case, arguing that because her children had been born out of wedlock, they were not eligible to recover damages. But when the case made its way to the Supreme Court, the decision was reversed. Arguing in favor of Louise's children, Justice William O. Douglas wrote that, despite their illegitimacy, the Levy children were "human, live, and have their being." These children, he continued, "were dependent upon her; she cared for them and nurtured them; they were indeed hers in the biological and in the spiritual sense; in her death they suffered wrong in the sense that any dependent would."[35] Later that same year, the court ruled similarly that the mother of a child born out of wedlock had the legal right to claim damages for that child's wrongful death.[36] And in a 1973 case, the court explicitly overturned state laws that had exempted men from bearing any financial responsibility for their "illegitimately" conceived children.[37] Legally, then, bastards no longer existed in the United States, and children born out of wedlock had the same relationship to their parents as anyone else.

In Europe, similar reforms passed from country to country, erupting first in Scandinavia and then spreading across the continent. Germany and the United Kingdom granted inheritance rights to out-of-wedlock children in 1969; France guaranteed equal legal rights to all children in 1973; and in 1975 a European convention recommended that all countries abolish discrimination between children born in and out of marriage.[38] By 1980, then, across Europe, North America, and China, one of the original slogans of the French Revolution had finally come to pass: the bastards were all gone.[39]

This shift in legal acceptance had wide ramifications—wider, by far, than was recognized at the time, even by the reformers themselves. Because in framing the arguments against illegitimacy in terms of the illegitimate child's best interests, the reformers had shoved a powerful wedge into the traditional legal structure of the family. Put simply: For centuries, the family existed and was structured largely to produce and protect children. The only way to do so, and the way that was thus enshrined by both law and custom, was to put a man and woman together in heterosexual union, and promise that their offspring—conceived through sex, affirmed by monogamy—would receive the fruits of their parents' labor. Children, therefore, were the channel through which their parents' wealth and property were preserved. The cases against bastardy changed all this, not just because bastardy itself was normalized but because the focus of legal protection moved from the family's interests to those of the child. This was the shift—subtle at the time, but crucial ever since—that has allowed subsequent family formations to be legalized, normalized, and eventually embraced. Because once the child is what matters, anything that produces and protects a child can be enshrined in both law and practice. Which is precisely what happened as assisted reproduction evolved.[40]

Love, Sex, and Contracts

In 1984, a New Jersey couple named Bill and Elizabeth Stern hired Mary Beth Whitehead to bear them a child. Bill was a biochemist and fully fertile; Elizabeth was a pediatrician who suffered from multiple sclerosis. Worried that the strain of pregnancy would be too much for Elizabeth's body, the Sterns had quietly advertised for a surrogate—a woman willing to conceive and bear their child. Mary Beth, a twenty-nine-year-old married mother of two, had answered their call. The Sterns agreed to pay Mary Beth ten thousand dollars and signed a surrogacy contract with her—one of the first known such contracts in the United States. She delivered the baby, received the money, and gave little Melissa to the Sterns.

After a few days, however, Mary Beth changed her mind. She decided that the baby should be hers, and determined to get her back. When Mr. Stern filed a complaint, Mary Beth seized the infant and fled with her, living for three months in a string of hotels and hiding places, and threatening to kill both herself and the child. Eventually, police in Florida forcibly removed Melissa from Mary Beth's care, and the case of "Baby M," as she became known, wound its way, slowly and painfully, through the New Jersey courts. In the end, the Supreme Court of the state granted custody to Bill Stern, based largely on what it perceived as the "best interests" of the child.[41] Melissa returned to the Sterns, grew up, and was later married by the judge who had presided over her trial.

The Baby M case was tragic, and not particularly high-tech. Bill Stern had impregnated Mary Beth Whitehead the almost-old-fashioned way, via artificial insemination. There was no IVF involved, no frozen eggs or PGD. But, in assigning parental rights to Bill, the New Jersey courts set what would become a powerful precedent: using the "interests of the child" to adjudicate between competing interests of several parents. It was exactly the same argument that Harry Krause had used to expand the rights of illegitimate kids—but with ramifications that reached considerably further.

Over the next ten years, the rising use of assisted reproductive technology forced courts in the United States and elsewhere to issue a series of landmark Solomon-like decisions, ruling time and again on questions that would once have been considered absurd. *Whose baby is it? And who gets, or has, to take care of it?* Because the state of California was increasingly home to the cutting-edge purveyors of the baby trade, California courts found themselves ruling in a disproportionate number of cases, and creating precedent for what would become a broader swath of law. In the high-profile 1990 case of *Johnson v. Calvert*, for example, a surrogate carrying the contracting couple's genetic child filed for custody of the baby.[42] Unlike the Baby M case, the surrogate here had a weaker claim to the child she had carried—it wasn't her genetic child, and the intended mother had provided the egg. And thus the courts, perhaps not surprisingly, gave full parental rights to the contracting parents. What was new, though, and crucial, was the logic of the courts' argument: "She who intended to procreate the child—that is, she who intended to bring about the birth of a child that she intended to raise as her own—is the natural mother."[43] In other words, parental rights came through intent rather than any natural biological link. Eight years later, the case of *Buzzanca*— stranger and sadder still—reinforced the 1990 decision and took it up a notch. In this case, a married couple named John and Luanne Buzzanca had used a sperm donor, an egg donor, and a gestational surrogate to conceive and create a child. Prior to the child's birth, however, John and Luanne divorced, and John disclaimed all parental responsibilities, claiming that he was not the child's father. The courts, however, found otherwise and, continuing the logic that now ran through *Baby M* and *Johnson*, ruled that because John and Luanne had initiated their baby's birth, they were her legal parents.[44]

If the law had stopped at this point—redefining parental rights to accord with contract rather than biology—it would already have been a massive shift, and a testament to technology's power to reshape the laws of nature. But things didn't stop there. Instead, a new cast of parents started to use the same laws and the same technologies to make an even more dramatic and ultimately successful claim: that same-sex parents should be treated exactly the same as their heterosexual counterparts.

The fight for marriage equality had begun much earlier, of course. After centuries of discrimination, persecution, and denial, gay men and women had finally become politically active in the 1960s, forming organizations such as the Gay Liberation Front and protesting, most famously at Manhattan's Stonewall Inn, against ongoing police raids of bars and nightclubs they frequented.[45] Much of their struggle was simply for acceptance—the right to be out at work, to embrace in public, to acknowledge a love and sexuality that had been banned for so long. Part of it, though, and a growing part over time, was for the legalization of same-sex unions, for the legal ability for two men or two women to enter the same kind of marital arrangements that had long bound and privileged heterosexual couples. In the early years of the struggle, most of the fight was led by lesbian couples, who had been forming families with the aid of artificial insemination since the 1970s.[46] They were subsequently joined by gay men, who focused—particularly during the height of the AIDS epidemic in the 1980s—on extending health benefits and hospital visitation privileges to same-sex partners. All these fights left their mark, and started to carve out both greater rights for same-sex couples and greater acceptance of their unions. But they didn't go nearly so far as to extend the rights of marriage to gay or lesbian couples.

What changed this equation—subtly but unequivocally—was assisted reproduction, and the extent to which these technologies had increasingly allowed gay couples to build families. Once again, it was lesbian women who had gone first, using basic technologies of artificial insemination to conceive children and become parents. Although the data here are extremely limited, solid anecdote suggests that lesbian women—usually portraying themselves as single—turned to commercial sperm banks almost from the moment of their inception. When I was interviewing these banks in the early 2000s, they generally reported that lesbian couples accounted for around one-third of their customers, with single women (who may or may not have been lesbian) accounting for an additional third. The children who resulted from these assisted conceptions were typically raised in lesbian households, where only one of the parents—the woman who bore the child—was legally considered its mother.

For gay men, obviously, the path to parenthood was considerably more circuitous, since they could not conceive a child without a woman's body and cooperation. Surrogacy opened the first window of possibility, but, especially in the wake of the Baby M debacle, it was hard to imagine courts ever granting rights to gay biological dads. Once gestational surrogates and donor eggs became options, though, the world changed for gay men. Suddenly they could imagine becoming fathers, using sperm from one partner (or, frequently, both partners mixed together, so that the genetic father remained unknown) to produce children they could call and claim as their own. Quietly, and then with increased fanfare, gay men began to flock to those parts of the world where surrogacy was legal (or at least not actively illegal) and their rights as contracted fathers were likely to be upheld. California was an obvious site, but other destinations—India, Nepal, Thailand, Mexico—entered the market, too. India alone boasted 350 clinics in 2009, many of which catered primarily, if not always explicitly, to a gay clientele.[47] Websites and facilitators then popped up around these centers, helping would-be dads navigate a medical and legal thicket that often involved one woman's eggs, another woman's womb, two men's sperm, and repeated rounds of IVF. The pregnancies and children that resulted from these blends were exceedingly expensive, costing anywhere from $40,000 to $120,000 in 2016.[48]

But for men who could afford the price tag, the combination was miraculous. Elton John became a father via surrogacy in 2010; Neil Patrick Harris had twins that same year, shared with his partner (later, husband) and delivered by surrogate.[49] Thousands more, less famous, followed a similar path to parenting. In many of these early cases, though, the two dads were legally confined to the same configuration that lesbian-led families had awkwardly settled into. One partner was the legal parent; the other was mom or dad in name only.[50]

And so, nearly as soon as they were established, such arrangements ran into trouble. Because what happened if the parents split, and the name-only dad or mom had no visitation rights? What if the legal parent died? Did his or her child become an orphan? Who got to sign the permission slip for school field trips, or had to pay for college tuition? As these cases made their way through the courts, a befuddled legal system increasingly turned to the few doctrines that were easy to apply: the interests of the child and the intentions of the would-be parents. In 2004, for example, a lesbian couple who had conceived two children via artificial insemination split up. One mother, who had given birth to the children, attempted to deny parental rights to her former partner, arguing that because the couple was unmarried, the other mother had no legal rights. But the courts found otherwise, ruling that because the couple "intended to produce a child that would be raised in their own home," both mothers were entitled to parental rights.[51] Similarly, in a 2005 case involving a non-biological lesbian mother who sought to avoid paying child care after she split from her partner, the courts found that the mother had already "held the children out as her own," and was thus a legal parent—even though she was in an unmarried relationship and had no biological connection to the children in question.[52]

In Spain and Sweden, similarly, the use of reproductive technologies by same-sex couples drove courts in the 1990s and early 2000s to gradually accept and then protect same-sex parents. In Sweden, partners in registered same-sex unions were permitted to jointly adopt a child in 2003, and were legally entitled to medically assisted insemination in 2005; in Spain, lesbian partners won the right to be treated essentially like husbands in a heterosexual relationship, with parenthood flowing naturally to them from the moment of the child's birth.[53]

From this point, the legal leap to marriage equality was relatively easy to make.[54] Because if the best interests of the child governed the aftermath of a couple's breakup or death, certainly it was relevant for happier moments as well. More specifically, if the best interest of a child was to be raised by stable, married parents, then the best interest of a child living with two moms or two dads was for those parents to be married as well. This was a central piece of the reasoning in *Obergefell v. Hodges*, the landmark 2015 case for marriage equality. Arguing that a "basis for protecting the right to marry is that it safeguards children and families," the U.S. Supreme Court held that same-sex couples, many of whom "provide loving and nurturing homes to their children," deserve equal rights to marry.[55] Similarly, in striking down California's short-lived ban on same-sex marriage, that state's supreme court focused explicitly on the rights and interests of the children involved. "A stable two-parent family relationship," the court argued, "is equally as important for the numerous children in California who are being raised by same-sex couples as for those children being raised by opposite-sex couples."[56] What the court didn't say was that the vast majority of these children had been conceived via assisted reproduction. They were the children of technology, and it was the circumstances of their births that helped propel their parents to marry.

Which isn't to say that the victory of marriage equality was driven entirely by technological forces. It came after decades of struggle by gay and lesbian activists, and after beloved celebrities like Billie Jean King and George Michael announced and embraced their homosexuality. It came on the heels of well-organized and well-funded political campaigns and in the shadow of the AIDS crisis. But it also came as a result of a fundamental shift in the mechanics of conception. Because once a sexual union between a man and a woman was no longer the only way of producing a child, other family structures became not only possible but inevitable. This is the world that assisted reproduction has wrought.

IVG and the Road to Poly Parenting

Nearly thirty years after the infamous Baby M case, another child named M was conceived. He was the desired child of a happily married heterosexual couple, and carried by the wife of this couple, who fully intended to mother him. Nearly old-fashioned, right? The only thing was that this Baby M's genetic mother was dead, killed by bowel cancer when she was only twenty-eight. Mrs. M, who won a court battle to conceive the child from her daughter's frozen eggs, was the baby's genetic grandmother.[57]

As technologies for assisted reproduction have rapidly advanced, stories such as these have become increasingly commonplace. There is the Indian woman who gave birth in 2016 to a son, conceived and carried when she was seventy-two. And the daughter who was artificially inseminated with her stepfather's sperm to give her mother another baby.[58] In areas of Japan, the government now regularly subsidizes egg freezing, urging young women to store and preserve their eggs for future pregnancies.[59] In Silicon Valley, tech firms increasingly include the costs of egg freezing as an employee benefit.[60] Recently, a friend described her sister-in-law's decision to conceive via IVF simply because her husband's company paid for the procedure. It was neater than sex, she figured, and more predictable.

Urged by technology, we have already changed how we procreate, and with whom. We have separated sex from reproduction, and multiplied the various pairings that can together produce a child. And we are about to enter a phase in which the underlying math—the number of partners, the pairing of genes—is itself up for grabs.

To understand the mechanics of in vitro gametogenesis (IVG), it's useful to return to the basic biology of conception. When humans reproduce—always, regardless of the person or position or sexual orientation—the man's sperm pierces the shell of the woman's egg, merging his genetic material with hers. The permutations of this mingling are infinite, carrying all the odds and intricacies that define an individual being. But the math is rigidly identical. Every sperm cell contains and contributes twenty-two chromosomes, plus the sex chromosome, which can be either an X or a Y. Every egg cell likewise contains and contributes twenty-two chromosomes plus the sex chromosome, which is always an X. When sperm and egg meet, the two pairs of chromosomes combine to form a blastocyst—a tiny bundle of cells that, when successfully implanted into a woman's uterus, will go on to become an embryo, and then a human. Crucially, each cell of the blastocyst contains forty-six chromosomes, half from the sperm and half from the egg.

Thus far, all the advances in assisted reproduction have been tweaks on or to this fundamental process. Artificial insemination brought the sperm toward the egg through a different, nonsexual channel. IVF mixed them together outside the woman's body. PGD enabled someone, other than God, to choose which clusters of egg-and-sperm would survive and prosper. Little things, really, in the broader sweep of life. What IVG does, by comparison, is to remove the fundamental link between one-man-one-sperm and one-woman-one-egg. Instead, it allows individuals to manufacture their gametes—their eggs or sperm—mixing and matching between genders and genes, and theoretically allowing more than two people to create a child together.

Here's how it works. Under natural conditions, the body produces eggs or sperm at puberty, taking undifferentiated stem cells (with forty-six chromosomes) and instructing them to split into more specialized cells, each containing just twenty-three chromosomes. In young men, the process occurs in the testicles, and these specialized cells become sperm. In women, it takes place in the ovaries, and the cells become eggs.[61] Both these processes are known as gametogenesis. In vitro gametogenesis, therefore, is precisely that: creating gametes not in the body but in the laboratory. More specifically, since around 2003 scientists have begun to find ways of coaxing human stem cells to produce eggs and sperm in a laboratory, to split the genetic material of an adult cell in half, in other words, and then configure each half—the newly created gamete—with the delicate precision that will enable its twenty-three chromosomes to eventually merge with another set to create an embryo.[62] To put it more bluntly: IVG can theoretically allow any adult to create an egg or sperm cell from a tiny sliver of their own skin.*

So far, IVG has worked only in mice. In 2003, separate teams of scientists in the United States and Japan figured out how to prod the embryonic stem cells of mice to develop into sperm or eggs. Already, by this point, scientists knew how to transform stem cells—the basic, "pluripotent" building blocks of an embryo—into bone or nerve or blood cells. But sperm and egg cells were tougher to produce, since the stem cell not only had to develop the precise capacities of a gamete but also had to split in half, dividing its forty-six chromosomes neatly into twenty-three. By tinkering with growth media, though, and then with finely tuned transcription factors (proteins), the researchers made it work: a team in Boston produced precursor sperm from the embryonic stem cells of mice; another in Philadelphia similarly produced precursor eggs.[63] Several years later, another team managed to mate the derived sperm cells with natural mouse egg cells, forming perfectly normal-seeming embryos. All the mice conceived this way apparently survived.[64]

Making the leap to humans will not necessarily be straightforward. Human cells are more complicated, and researchers are understandably wary of the laboratory repetition—hundreds of cells, thousands of trials—that a viable human embryo would entail. In the United States, such research might even be deemed illegal.[65] But, with very few exceptions, recent history suggests that advances in reproductive

---

* The adult cell would first need to be "de-differentiated"—essentially, converted back into what is called an induced pluripotent cell.

technologies nearly always leap eventually from the animal world to humans.[66] If we can figure out how to make babies, and to configure their creation in more precise ways, we do it. We did it with IVF, despite howls of bioethical criticism; we did it with PGD; and we are likely to do it again with IVG.

As techniques of IVG move into the medical mainstream, therefore, we will start to see families creating gametes first, rather than children or embryos. A single woman, for example, might mix her own egg with sperm fashioned from the genetic material of her two best male friends. Or her two best female friends.* She might mate her egg with a carefully selected donor sperm, using PGD to eliminate any risk of the cystic fibrosis that runs in her own family. The resulting embryo could then yield a next-generation egg to be paired with her best friends' similarly well-conceived sperm. And so on.

Note what happens in any of these cases. First, and most obviously, IVG carries us as a species ever further down the road of genetic selection. It magnifies the power of IVF and PGD, allowing individuals to handpick and manipulate not only their child's genetic parents but those parents' parents as well. As humans, it brings us that much closer to playing God.

The second implication, though subtler, is perhaps greater still. And it concerns not the children born of IVG, but the parents who conceive them.

This is the part that brings us back to the math, and to the fundamental ways in which IVG reconfigures both the biology of reproduction and the social norms that have long surrounded procreation. Ever since humans began to evolve, the seeds of our creation have lain in mated pairs of diploid cells—one egg, one sperm, each hatched in the body of a postpubescent adult and containing twenty-three chromosomes. Ever since we began to link our procreative patterns to our

---

* In this case, one of the women's eggs would first need to be matched with a sperm, creating an embryo that could then be used to derive a next-generation male gamete. Currently, there is no way of creating the crucial Y chromosome from women's genetic material alone.

social ones—ever since, that is, the dawn of the Neolithic Age—we have embraced two-person heterosexual pairings as the inviolable structure of parenting. One man, one woman, forty-six chromosomes, and a child. IVG, though, violates this structure. Blows it up entirely, in fact. Because by separating the gametes from the individual, IVG removes the two-parent requirement of reproduction, allowing for an entirely new range of possibilities. If our hypothetical single woman above chooses to reproduce with her two best friends, the child that results has three genetic parents. If she instead creates an embryo using donor sperm, generates a new egg cell, and mates it with a sperm cell that two friends have produced, the child has four parents. And so on. If the revolution of IVF was to liberate reproduction from sex, then the even bigger revolution of IVG is to dismantle the reproductive structure of heterosexuality. Once upon a time, defenders of heterosexual marriage could easily and convincingly argue that "marriage is intrinsically a sexual union of husband and wife, because these are the only unions that can make a new life."[67] With IVG, that's no longer true. Instead, two men can make a baby. Two women can make a baby (with a modicum of outside assistance). Four sexually unconnected housemates can make a baby. And that changes everything we've ever known about sex and babies and marriage.

In the early days of IVG's adoption, the most obvious users of the technology are likely to be gay couples. Instead of having to purchase donor eggs, men like AJ will be able to use their own skin cells—and thus their own DNA—to create egg cells that can then be paired with their partner's sperm. Lesbian women will have to endure a few more complications in the process, but they will be able to pair one woman's egg with a sperm derived from another woman's genetic material.[68] For the first time in history, same-sex couples will be able to conceive children who are wholly and genetically "theirs."[69] Over time, however, interest in IVG is likely to seep—as reproductive technologies always do—into the broader population. Single women will inevitably want to use it, either with friends or family, or with their own genetic material.[70]

Platonic friends may choose to become parents together, sharing a life and a family that are not linked to sex. Older couples could conceive and raise their own grandchildren. And in a possibility raised long ago by both utopian and dystopian literature, communities of like-minded people could manufacture their own descendants, protecting a lineage that would otherwise die from lack of sex or interest.[71]

Already, legal and scientific observers have started to ruminate on these possibilities, generally considering them under the umbrella term of "multiplex parenting."[72] I don't like this term. It's vaguely reminiscent of a real estate development, to begin with, and it threatens to define those who will eventually become parents through IVG as members of a small and highly specialized state. I would prefer and suggest instead "poly parenting." It's a nicer phrase, it rolls off the tongue more easily, and it's swiftly reminiscent of related social patterns—polyamory, for example, and polygamy. Because that, after all, is where poly parenting will inevitably take us.

Specifically, once we start imagining, and then living in, a world of fluid parenting, then it's just a hop, skip, and legal nicety to undo or at least revise our centuries-old conviction that procreative unions—like Noah's animals—come only in pairs. Maybe our species' new ark is composed of a more motley crew, of threesomes and foursomes, old and young, men and women and those in between, reproducing with whomever they choose and loving as they desire.[73] IVG alone, of course, can't create that world. And it will take a long time to dismantle norms of marriage and parenting that have been around since, well, Noah. But the history of assisted reproduction is powerful and clear: once we create new technologies for conception, we embrace them. We worry, always, about mucking with nature. We fret about designer babies or the possibility of some madman hatching Frankenstein in his backyard.

Then we discover that it's just the nice couple next door, using an increasingly common technology to create children they love, children who become far less scary as they move out of the world of scientific abstraction and into the realm of the real. Over a remarkably short period of time, we have grown accustomed to those nice parents in our neighborhood being a couple of men, or women, or a single one of either sex. We will get used to them being a threesome or foursome as well. And then we'll see the new normal as simply the real, and we'll forget that it was technology that changed this world.

REMEMBER THE JETSONS? They were the cartoon family in a popular 1960s television show—a far-off-in-the-future family, living a streamlined and utopian life. Their maid, Rosie, was a robot. Their dog, Astro, talked. They lived in a floating apartment in Orbit City and zipped around in an airborne car that looked distinctly like a flying saucer. What *The Jetsons* got fundamentally wrong, though, was the *family*. The Jetsons themselves: George, the bumbling but well-intentioned father; Jane, his elegant stay-at-home wife; and Elroy and Judy, their forward-looking but otherwise completely standard 1960s middle-class kids.* The technology around them—flying cars, dinners delivered by pill—had changed profoundly, but they stayed exactly the same. One man. One woman. Two healthy children who were hatched, presumably, by old-fashioned means. Which is, in many ways, precisely the opposite of what's actually occurred. Over the past fifty years, the technology of reproduction has advanced further and more quickly than nearly any other sphere of production. We are still driving to work pretty much the way we did in the 1960s. We eat our meals the same way. But we are reproducing in ways that would have been truly unimaginable just several decades ago: Two men and a surrogate. Two women and a sperm donor. A man, a woman, and a high school acquaintance. A single person and their own manipulated genes, twisted and selected to produce children of their own.

As each of these permutations emerged, popular concern was fo-

---

* One of the futuristic gadgets was so prescient as to seem nearly eerie now—DiDi, the digital diary that talks with Judy for hours each night.

cused, always, on the children. Would they be normal after all this intervention? Would they be stunted by their laboratory origins, or forebears of a better race? We have several generations now of these once-newfangled babies; millions of children born from test tubes and petri dishes and frozen Ivy League eggs. And the overwhelming verdict is that the kids are all right. More than all right, actually. They're just— well, normal. Far from being superkids or signs of their fathers' progressive parenting stance, Josh and Zoe are just Josh and Zoe, hanging out in the living room and carving their own paths through life.

What has changed fundamentally, though, is the nature of parenting and families. The old world of George and Jane Jetson, of man-and-wife and babies produced in candlelit bedrooms, is going the way of the stork. Not all at once, of course, and not for everyone. But the technologies of assisted reproduction have pushed a powerful wedge between sex and procreation, and thus between the structure of the family and its function. In other words, once science shattered the link between sex and reproduction, it sundered the ancient connection between marriage and family. As reproductive science expands, which it inevitably will— and as family structures morph and multiply, which they inevitably shall—the dire predictions of many social conservatives may well come to pass. Marriage as we know it may not survive. Except it won't be liberals or homosexuals who destroy it. It will be science.

5.

Sex and Love Online

We are in uncharted territory . . . The first [major transition in heterosexual mating] was around 10,000 to 15,000 years ago, in the agricultural revolution . . . The second major transition is with the rise of the Internet.

—Justin Garcia, Kinsey Institute for Research in Sex, Gender, and Reproduction, 2015[1]

Tinder is a scary place, at least for those of us who came of age at a time when one dated and generally had sex with people one already knew. On Tinder, though, friendship isn't really a thing. Neither are family connections, hobbies, or silly predilections for piña coladas and getting caught in the rain. Instead, Tinder is just about sex—straightforward, fairly anonymous, often pretty-near-instantaneous sex. You sign on, fill in the most basic information (age, education, location, and the gender you're interested in), post a photo or two, and then you're off to the races.

Launched in 2012 by two former students from the University of Southern California, Tinder uses touch technologies (no pun intended) to make the process of partnering (or at least the process of *finding* sexual partners) as easy and painless as possible. Rather than searching for mates from a list of preselected, algorithmically optimal picks (the model on earlier sites like Match.com or OkCupid), rather than wandering to a bar or club, rather than—heaven forbid—asking your roommate's cousin's friend to set you up with his brother, Tinder lets you just swipe. That's it. See a bunch of photos. Review an apparently infinite reel of perpetually available men or women (or both) and just swipe. The left swipes disappear into the digital dustbin; the rights stay put, offering an instant opportunity for texting, or sexting, or sex. As of 2018, Tinder reported 50 million users each month, nearly all of whom, in theory at least, logged on to the site to find sex, or at least the beginning of what might be a relationship. Grindr, Tinder's gay precursor, boasted nearly 5 million active users, while niche markets—finding Chinese singles in Dubai, say, or live-chatting with potential Russian brides—were proliferating across the Internet and around the world.

It is easy to deride these sites, or to giggle somewhat prudishly at their explicit offerings. It is tempting, always, for one generation to bemoan the sexual practices and preferences of the next. But there is something fundamentally important happening on sites like Tinder and Bumble and Hinge, something that promises to change not only how young people manage their social and sexual encounters but also, and more important, how they build their lives.

Throughout most of history, recall, sex (particularly for women) came tightly packaged with both marriage and children. Contraception and assisted reproduction severed that link, allowing for sex without children and children without sex. Now innovations like Tinder are ripping the old connections even further apart, offering and delivering sex without even the hint of an ongoing relationship. This isn't prostitution, where women exchange sex for money (though that, too, is available online). Instead, it's simply sex for sex, mediated and facilitated by incredible computing power.

So what happens when sex—and love, and lust—move into the virtual world? What happens when people find each other through photos and swipes rather than church socials or freshman English class?

One might assume that young people would fall into a surge of sexual relationships—that, freed from the clunky constraints of social norms and commitments, they would surf from one no-strings-attached hookup to the next. One might assume that these technologies would give young people an almost unlimited buffet of sexual offerings—willing partners, easy and available, only a phone call away.* At the risk of being blunt, one might presume that the users of Tinder (and Grindr and Hinge) were having lots of sex, with lots of people, in configurations that ignite just a hint of jealousy from those unlucky enough to have been born in the age of the rotary phone. If one paused to think about it, one might also presume that a generation raised on sexting and swiping would still, eventually, graduate to dating's last stage: to marriage and commitment and kids.

But they're not.

Instead, even though it's still a little too early to tell, current indications are that the members of the Tinder generation are not settling into the same patterns of dating and marriage that defined—and confined—their parents. They are not getting married at anything close to the same rates. They are not having as many children. And, most shockingly of all, they are not having as much sex. In fact, in 2015, only 41 percent of American high school students reported having had sex (down from 54 percent in 1991) and 15 percent of people ages twenty to twenty-four reported having had no sexual partners at all since the age of eighteen.[2] In the United Kingdom, the frequency of sexual relations fell by roughly

---

* Or a text, more precisely. For the Tinder generation, phone calls are rapidly becoming an antiquated technology.

20 percent between 2000 and 2013; in Japan, fully 43 percent of single people between the ages of eighteen and thirty-four were virgins in 2015.[3]

Some of these trends may be linked to a whole range of other developments—a slower economy, for example, fear of the future, or a desire to postpone parenting until later in life.[4] But some, almost certainly, are linked to the patterns of romance brought forth by Tinder.

In particular, the infinity of choices that proliferate on sites like Tinder seems to be generating a strangely bimodal pattern of sexual behavior. Some people—the beautiful, the witty, the ones with good teeth and sparkly profiles—can flit from relationship to relationship, swiping right for a quick encounter, and disappearing ("ghosting") well in advance of anything that might look or feel like romance. Others fall through the gaps of the app, dismissed by people they never even had a chance to meet. Some people are indeed having lots of sexual encounters—albeit with mixed levels of satisfaction. And some are having virtually none at all. In Japan, for example, young people are frequently resorting to what has been dubbed "Pot Noodle love"—instant gratification via online pornography or virtual sex, rather than actual trysts with actual humans. Again, several causes might explain this behavior. Online pornography is ubiquitous and increasingly varied. Virtual sex—with avatars or chatbots or specially built robots—is becoming easier to imagine and obtain. But Tinder and its peers seem clearly to be driving these trends as well. Because, strangely perhaps, and counterintuitively, having so many theoretical options for real-life encounters is upping the bar on making them work—upping it so high that many people are pulling away, preferring the dull thud of loneliness to the sting of rejection.

As these trends make their way across the demographic landscape, they are almost certain to reshape the contours of both love and marriage. In fact, they already are: in 2012, only 22 percent of American men and 33 percent of American women were married before they turned twenty-five, down from 72 and 87 percent, respectively, in the early 1960s.[5] In 2011, a whopping 48 percent of Europeans between the ages of eighteen and thirty still lived with their parents. And in Japan, a third of people under thirty have never dated at all.[6]

Again, it's not just Tinder or dating apps more generally that are causing these shifts, but they are almost certainly a major contributor, a technological innovation that is already disrupting how an entire generation is meeting and mating and falling in love. In a span of less than twenty years, online algorithms have revolutionized the dating process, prodding their still mostly millennial users to build relationships in ways that are fundamentally different from those their parents and grandparents experienced. As these technologies evolve and proliferate, they are likely to have an even more profound effect, changing not just how we meet potential partners but how we interact with them as well. Because once we grow accustomed to swiping left or right—to making intimate decisions in a split second, with limited information and selective visual cues—we will almost certainly yearn for other relationships that are similarly fast and fickle. For jobs and friendships that can be easily replaced, for example. Preferences that shift according to others' likes and dislikes. For unions more ephemeral than marriage, and sex divorced from context. Push these technologies to their logical extreme, and you can even start to imagine how relationships of the future could return full circle to resemble those of the ancient past—back to the days before private property and industrialization carved our commitment to marriage and monogamy.

And it will have been Tinder—and JSwipe and OkCupid and Bumble—that made us do it.

History of the Swipe

Until very recently, marriage was a community sport, with matches proposed and consummated within a strictly defined social group. Men and women married members of their own tribe—be they Australian aborigines, Jewish shtetl dwellers, or the British aristocracy—and followed well-defined norms of courtship. In many cases, older intermediaries brought the young people together, either with or without their consent. Jewish communities, for example, relied famously on *shadchanim* (often mistakenly referred to as *yentas*), who knew all the families in a given town and tried to match their offspring in appropriate pairs.[7] Villages in India had local matchmakers who carried résumés door to door; traditional Chinese marriages were arranged by parents, who followed the rule of "matching doors and parallel windows" and aimed to pair their children with those from similar backgrounds.[8] Romance, much less sexual chemistry, was rarely considered part of the process.

Things started to change in the nineteenth century, prompted once again by broader technological shifts. As the Industrial Revolution swept across the world, it created the mobility of both place and opportunity described in chapter 2. New modes of transportation gave young people the opportunity to leave their homes and villages. New kinds of jobs—in factories, on railroads, and in burgeoning urban centers—allowed them to earn wages that did not rely on either their families' traditional occupations or, necessarily, their approval. So the son of a Lancashire farmer could travel to London and become a cabinetmaker for the king.[9] The daughter of a Jewish milkman from a tiny Belorussian town could dream of following her true love to Siberia.[10] Young people—older people, even—began to imagine meeting and mating beyond the boundaries of their ancestral communities. Which created both problems and opportunities.

The main problem was that people who moved into new communities lost the social structures of their old ones. The *shadchanim* disappeared; the debutantes abandoned their balls. With few established channels for meeting potential mates, or even learning of their existence, men and women far from home turned to new ways: to bars and social dances, newspaper personals and matchmaking correspondence groups.

At their core, all these methods addressed the same two problems that animate Tinder today: how to find people, and how to select them. The first is a problem of information, the second a matter of preference. Think for a moment of how a *shadchan* or any traditional matchmaker behaved. She knew all the young men and women in a given town (the information). And she knew, roughly at least, who was eligible to be paired with whom, given family status, social standing, or whatever norms dominated in her place and time (preferences). It was a limited pool of data, but she had it all.

Once the *shadchanim* disappeared, however, men and women in search of mates or dates had to acquire these two bodies of information on their own. *Who was available? And who was of interest to them?* Newspaper personals (which were never more than marginally popular) solved a piece of the first puzzle, allowing individuals to broadcast their availability to one another. But they did little to solve the preference piece. Social gatherings geared to certain demographics—college mixers, church socials, community youth groups—put groups of apparently similar people together but did little, again, to address preferences. Not that this was a huge problem. Most people coming of age during the college mixer/church social era did somehow manage to stumble their way toward a mate, perhaps even a lifelong love. Across the industrialized world, in fact, most men and women in 1950 were married before they turned twenty-five.[11]

The advent of the digital era, however, tempted a new generation of young people to apply the technologies of computing to the algorithms of romance. As early as 1959, when computers themselves were still in their prepubescent phase, two electrical engineering students at Stanford University created a "Happy Families Planning Service" for their Math 139 class. The idea was straightforward: to use Stanford's first large computer, the IBM 650, to match one set of forty-nine items—undergraduate men—with another set of forty-nine items—women. The student designers, Jim Harvey and Phil Fialer, presumably wanted to learn the language of programming. But, having just recently been punished for broadcasting an inappropriately sexual show on the campus radio station, the pair also wanted to have some fun. So they threw together a questionnaire, with queries that ranged from respondents' height and weight to their hobbies and personal habits. Then they hand-graded each question (along probably dubious criteria) and calculated a score. The computer came in only at the very end, calculating a "difference score" for each possible male-female pair. The math was easy, the psychology suspect. When the professor graded the project without actually running the program (each project had been assigned only ten minutes of computing time), Harvey and Fialer broke into the computer rooms late one night and ran it themselves. Nine hours later, they had their results. The students received an A on their project, and at least one known marriage resulted.[12]

The next year, undergraduates at Harvard University launched a similar project, reaching out this time to college students across the United States, who paid three dollars for the privilege of having their personal information fed into an Avco 1790 computer. In the United Kingdom, two erstwhile marriage bureaus merged in the mid-1960s to form Com-Pat, a computer-aided service that paired single women with both marriage prospects and male escorts.[13] And in 1965, a company named Technical Automated Compatibility Testing (TACT) opened New York City's first computer-dating service, with print advertisements that, somewhat predictably, featured photographs of beautiful blond women.[14] Matches, when they occurred, arrived the old-fashioned way: on paper, via mail.

For about thirty years, computer-mediated dating remained in what might generously be termed a niche market. There weren't many customers. There weren't many marriages, or even particularly good dates. Instead, a slightly seamy pall hung over the practice, hinting at women on the verge of spinsterhood and men with something to hide. There were video offshoots that emerged in the 1980s, and 900-number phone providers, but nothing that affected the broader norms of dating and mating and sex. Until 1995, when Match.com was launched.

Match was the brainchild of Gary Kremen, another Stanford graduate, who (a) understood the potential of disruptive technology and (b) was having a hard time finding a date. Like many of his Silicon Valley peers at the time, Kremen was searching for something to do with the Internet, some way of harnessing its speed and power and reach. Most of his peers were focused on business opportunities for this breakthrough technology, on what were then categorized as either business-to-business (B2B) or business-to-consumer (B2C) applications. Kremen's insight was to link individuals instead, using the vast reach of the Internet to perform a startlingly primitive function: introducing one single human being to another.[15] Specifically, for $9.95 a month, Match.com offered to introduce its subscribers to a select group of potential mates—people in their neighborhood who were looking to connect and who had similar interests or preferences.[16] Users would simply sign on, complete personal profiles, and then request to match with other members. It was an ancient exchange, one person learning about another and looking for a spark. What Match added to the equation, though, were the crucial assets of size, speed, and information. Size, because cyberspace potentially included a mind-boggling number of human beings, any one of whom could be a perfect mate. Speed, because it brought these humans together in the blink of an eye. And information, because apps like Match allowed these vast pools of people to be sorted and preselected, gathered into pairings that, at least in theory, made sense. The back-page personals had become the world's biggest market—global in scope and effectively infinite.[17]

The next step in this evolution arose from a strange but powerful confluence of gay men and mobile phones. In 2008, Apple released the second version of its iPhone—one of the most successful product launches of all time. Within three days, it had sold an astounding one million units, and it spawned an even greater number of copycat phones.[18] Even more important, Apple had timed the phone's release to coincide with the opening of its App Store—essentially, a platform on which any independent developer could build a software application (app) to run on Apple's phone. Almost at once, the market took off. People built apps to search for hotel rooms and listen to music. They built apps for mobile banking and restaurant reservations and college admissions. Indeed, some analysts have described the wave of app-building that followed the iPhone launch as a revolution in its own right, a "cultural, social and economic phenomenon that changed how people work, play, meet [and] travel."[19] By June 2018, developers had earned more than $100 billion from products sold through the App Store.[20]

One of the most successful of these early entrants was an app called Grindr, founded by a young gay entrepreneur named Joel Simkhai, which brilliantly combined the iPhone's visual interface with its global positioning capabilities. Simply put, any iPhone by this point could take and store thousands of photos. Any phone could instantly track and re-port its user's location. Grindr bundled these two functions into a dis-creet and easy-to-use service for gay men, allowing them, at any time or place, to view photos of dozens of nearby strangers. If they liked, they clicked, and could instantly receive the man's personal profile and pre-cise location. In its first year of operation, Grindr registered over 3.5 million users in 192 countries.[21]

At first, Grindr elicited little more than titillation from the straight-dating community. But entrepreneurs quickly started to realize that its mix of photos and instant contact could work just as well with hetero-sexual users. In 2012, two friends named Sean Rad and Justin Mateen thus launched Tinder, riffing upon Grindr's underlying model of match-ing: quick, spontaneous, visual.[22] There are none of the questionnaires that fueled earlier dating services, no hidden formulas for matching per-sonality types or political preferences. Instead, Tinder users simply cre-ate a brief profile from their existing Facebook accounts. They include a modicum of personal information—age, location, and the preferred gender of their partners—and upload a handful of photos. Almost in-stantly, they are then presented with a constant scroll of potential partners—photos framed as playing cards that demand an immediate reply: swipe right for yes, left for no. If both users swipe right, a pop-up animation links their photos and allows for direct messaging. It's easy. It's painless. It's virtually anonymous until the point when a meeting is arranged. "The Twitter of dating," according to one relationship coach. Or simply "McDonald's for sex."[23]

Like Grindr, Tinder took off nearly from its outset, registering 10 million daily active users by 2014.[24] Other dating sites (JSwipe, Bumble, Hinge, Coffee Meets Bagel) essentially adopted its quick-and-visual model; other behaviors (choosing restaurants, rating movies, selecting political candidates) were reshaped by the "swiping left" model. Yet interestingly, and despite rumors to the contrary, Tinder doesn't seem to have created, or even abetted, an epidemic of casual sex. In fact, according to one 2017 Dutch study, fewer than 20 percent of active Tinder users had ever used the app for a one-night stand, and more than 25 percent had used it to forge what eventually became a committed relationship.[25] But still. Even if Tinder hasn't wholly upended contemporary patterns of sexual behavior, it and the other online dating services have incontrovertibly transformed the norms of attraction, changing—perhaps forever—the process by which people seek, find, and fall in love with one another.

In the time before Tinder, people met in person and across existing social networks. They generally learned to like, maybe even love, each other before having sex. After Tinder, people learned to meet online instead, devoid of social context or constraints. They met more quickly, decided more quickly, and had the ever-present option of returning to scroll and swipe again. These are the changes—much more than hookup sex—that Tinder and its ilk have wrought. Broader social networks. Lightning-quick decisions. And an infinity of romantic choice. Put these together, and the age-old game of sexual attraction begins to be governed by a different set of rules.

Love in the Time of Instagram

Usually, I don't feel obligated to experience the things I'm writing about. I didn't buy diamonds when I first wrote about that market in the early 1990s; I didn't undergo IVF when I wrote about reproductive technologies in 2006. But, although I'm happily married and a little bit reluctant to drag any portion of my private life onto the Internet, it didn't seem right to be talking about the intricacies of online dating without at least dipping a toe into the water.

Me on Match

So I registered on Match.com and Tinder, more or less honestly and with a Bitmoji for my photo.

Here's how it went: Match.com gently takes me through a fifteen-step process of defining who I am and what I want. Middle-aged woman seeks middle-aged man for something that could range from "just browsing" to "long-term relationship." I list my height and weight and body type, chart my education level and marital status. (I have to lie on that one. Match.com has no provision for married folks looking to browse.) By the time I hit number 11—Religion—small photos of indeterminate men start marching across the bottom of my screen. (One is drinking a beer, another stroking a cat. I have no idea why. Given a choice of interests, I have selected neither beer nor cats.) I write the shortest blurb the site will allow: "I like coffee, chocolate, good books, and grammar." Not exactly prime dating fodder.

But the men stream in nevertheless. Luckyrusty is first, followed by John and Nightflight747. There's Spanky (I don't even want to pause there) and LoverClaudio. When I hover briefly on Bimmerboy (who, with glasses and a tie, actually looks like someone I could encounter in real life), Match prods me to discover "More like him!" So I leave Bimmerboy and am swept over to Handyman123—no tie, "enjoys laughing," no obvious connection to Bimmerboy's more serious stance. But who knows, maybe he likes grammar, too.

Within an hour, and then repeatedly over the next few days, my inbox is flooded with men. Nice-looking men. Decent-seeming men. Mostly divorced, mostly dads, mostly (at least in theory) willing to meet a self-professed middle-aged woman with nothing of interest to say and a cartoon for a photo. The strangest part, though, are the come-ons that are clearly being generated by the site itself. "He just emailed you!" Match.com reports breathlessly. "Could he be the one?" Clicking on these entreaties does what you'd imagine: it takes you to the page of progressively more expensive subscription plans. Still, I must confess, there is something vaguely nice, even intriguing, about the attention. *He likes me!* A little jolt of dopamine to get you through the day.

Tinder is much more direct. Fewer words, a handful of photos, a parade of available men marching across my phone at the moment I desire them, or sex, or even just a fleeting fantasy of what might be.

My eye was still the voyeur's, though, and it was almost certainly tainted by both (a) my age and (b) the fact that I wasn't doing this for real. And so, to get a better sense of the online experience, I convened a series of informal conversations during the summer and fall of 2018. The rules of the gatherings were simple: I provided Greek takeout, plenty of wine, and a promise of anonymity. The guests agreed to tell me their Internet dating stories. It wasn't a scientific sample, or perhaps even a particularly representative one. They were mostly people in their twenties and thirties, more educated and urban than average, heavily weighted toward straight and white and female. Still, their stories were fascinating, and lined up closely with what more wide-ranging and quantitative data suggest. Part of what they related felt sweetly eternal to me: the pings of adrenaline, the shock waves of love, the lure of lust, and the heartbreak of rejection. All of that is occurring across Tinder and its online peers; much of it, in fact, is merely being amplified and broadened by technology's reach. But part of what they're experiencing online—part, even, of what I glimpsed through my halfhearted rummaging—is fundamentally different. The sheer range of options. The ability to experiment with sex and sexualities that fall far outside what was once the norm. And the possibility that choice itself can erode the contours of commitment.

Let's begin with what hasn't changed: the pull of attraction itself. When people sign on to any of the online dating apps, they are looking to connect—for coffee or sex or marriage or whatever—with someone who sparks an interest in them. Someone who matches. Someone special.

Some of the sites—Match.com, most prominently, and eHarmony—base their business models on being smarter than the users themselves, on employing a combination of profiles and algorithms to select relationships that the parties themselves might never stumble upon. Others, like Tinder, stay out of the matchmaking game, letting users simply scroll and search until someone piques their interest. In either case, though, people are looking for what remains deliciously unquantifiable: attraction. As Helen Fisher, the anthropologist who also serves as chief science adviser to Match.com, says, "Thirst and hunger are not going away just because you swipe left or right on Tinder. It's the same thing for romantic love . . . the ancient brain clicks in just the way it always has."[26] Certainly, this seems to be the experience of the millennials with whom I spoke.

What also doesn't change online is the emotional content of attraction; the inevitable highs and lows of being in love and lust. They're just being channeled through different devices. No one on Tinder will ever sit forlornly by the kitchen phone, willing it to ring. No one will have bundles of lovesick letters stored in attic shoeboxes. But they do have the distinctly digital replicas of these same behaviors. The smartphone they sleep with, waiting for a buzz on the other end. The emojis they carefully curate, trying for that precise combination of flirtation and remove. When I asked my dinner guests how they remember their romantic events, every one of them quickly retrieved digital records: screenshots, usually, of that first interaction. "I can't recall anything," says Jade, a petite human rights lawyer, "that I have a stronger emotional connection to." Similarly, when people decide to be together and exclusive, they frequently mark the event by mutually deleting their online profiles.

Finally, although online daters are bombarded each day with a multitude of romantic options, they seem to fall—over time and in the aggregate—into certain predictable patterns. Some women regularly select for tall, dark, overeducated engineers. Others for volleyball players, or middle-aged guys who remind them of Che Guevara. Some men lean toward mother figures, or curly hair, or particular body types. Time and again, my dinner guests reported, they would scroll and scroll and scroll—and find themselves drawn to someone who was disarmingly similar to those who had attracted them before. In this regard, online dating revolutionizes the information part of mating—by providing so many, many options—but keeps the selection part oddly unchanged. Even without *shadchanim* or intervening parents, it seems, people self-select for certain, eventually predictable, types.[27]

So what, then, is different? It's subtle, but a lot. To begin with, the oft-repeated rumor/fear/promise of sex-before-dating is decidedly, if not universally, true. Especially on sites like Tinder and Bumble, which are explicitly premised on quick and noncommittal interaction, people find each other, get physical if they so desire, and *then* decide whether the spark of attraction warrants a date or longer-term relationship. Technology isn't solely responsible for this changing state of affairs (pun intended), but it has been a major driver. Because once people can reliably expect to have a wide array of potential partners—much greater than anything a village, or church dance, or freshman mixer could ever have produced—they don't need to make any commitment before having a drink, a kiss, or a one-night fling.[28] Moreover, because they know (or at least strongly suspect) that any partner feels similarly, they explicitly don't want to announce any sense of commitment, or even attraction, at the outset. That comes later, if at all.

Meanwhile, although the users of online dating apps tend to click repeatedly on similar types of people (the tall, dark, handsome engineer phenomenon), they also get exposed to geographies and experiences that are far broader than anything they might reliably encounter in the real world.* During one of my informal gatherings, two white people mentioned that they had met and dated people of color, something that they had not done when dating only within their preexisting social circles. An African American woman met mostly African American men in her "real" life and mostly white men online. Two women first experimented with lesbian relationships online; one man encountered, and

---

* IRL, or *in real life*, as my dinner guests gently reminded me.

subsequently embraced, polyamory. All felt certain that they wouldn't have discovered these circles or people without their online apps. Or as Mark,[29] newly polyamorous, explained: "I never would have thought about being nonmonogamous unless the app asked me." For all my guests, online apps had given them information and confidence that would have been harder to come by in a bar or other social setting. They provided privacy and room for experimentation: "They open you up," one thirty-five-year-old woman agreed, "to all kinds of possibilities." And, already, the prevalence of these apps has coincided with a notable increase in interracial marriage and relationships: after the launch of Match.com in 1995, the frequency of interracial marriage in the United States grew rapidly; following Tinder's launch in 2014, it jumped upward again.[30] Although it is still too early for the data to be definitive, a careful 2018 study concluded that "online dating is causing more interracial marriages, and that this change is ongoing."[31] Yes, the authors acknowledge, the racial composition of the United States is changing as well, but not nearly at a rate sufficient to account for the recent surge in intermarriage.[32]

Between 2000 and 2015, the percentage of newlyweds who married across racial lines in the United States leapt from 10.7 to 17—an increase of nearly 60 percent.[33] Between 2001 and 2011, it rose a similar 35 percent in the United Kingdom.[34]

Meanwhile, in cultures that have until recently hewn to older models of arranged marriage, dating apps have given young people more avenues through which to defy society's norms and their parents' wishes. In India, for example, websites like TwoMangoes and TrulyMadly offer men and women a chance to meet and socialize, to find "really cool people all around the world," as TwoMangoes advertises, without the pressure to commit to an early and socially approved marriage.[35] Young people—especially the affluent and more educated—are increasingly posting their own profiles online (rather than ceding that task to their parents) and assessing potential partners before agreeing to meet them. "Twenty years ago," said the founder of one well-used marriage site, "parents chose the matches . . . Now the brides and grooms are in the driver's seat."[36] In China, dating sites Jiayuan and Baihe registered 126 million and 85 million users, respectively, in 2015; in Saudi Arabia, whose deeply conservative mores prohibit a woman from being in close physical proximity with any unrelated man, young people are increasingly finding each other online, exchanging texts and messages and the occasional unveiled photograph.[37] Even on more traditional marriage sites, Saudis can now view thousands of potential spouses—and learn more about them than would likely have been the case in real-life courtship.[38]

What happens to us as humans, though, when we're confronted by so much choice? We know, from social science experiments, how it goes for jam: If shoppers are presented with six new varieties, attractively displayed and well priced, they will tend, on average, to splurge, and purchase a jam to bring home. But if there are, say, twelve varieties of jam, equally attractive and priced—from boysenberry-champagne jam to red pepper jelly to mango-peach chutney—the average shopper will skip buying jam altogether. She'll just walk on by, retreating into the safer aisles of peanut butter, or maybe even deciding that she's not hungry at all.[39] That's what too many jam choices do for you. Or too many channels on the television. And maybe, it seems, too many dating options as well. Faced with so much choice—too much choice—people just retreat.[40] As my friend Phoebe, a twenty-nine-year-old playwright, confided recently: "The worst part about this new age of digital dating is that I can no longer complain about dating. I can no longer bemoan the fact that it's impossible to meet someone in New York City, because as soon as I try to vent about the tribulations of dating I'm cut off by a phone shoved in my face, displaying the latest dating app, which literally puts thousands of single men (in my zip code, no less) at my fingertips." She goes on: "I know fairy tales aren't perfect, but Cinderella had her glass slippers and her one and only Prince Charming, while I'm stuck with JSwipe, where my 'one and only' is hundreds of Jewish doctors, lawyers, and accountants who think citing *Catcher in the Rye* as their favorite book makes them seem smart and mysterious."[41]

Moreover, even once people plow through the choices buzzing at them from their laptops and smartphones; even once they have clicked and chosen, clicked and matched; even once they have *found the one*—there are still so many others out there, muddying the question of commitment with the addiction of choice. Camilla, for instance, a twenty-nine-year-old publishing executive, has been in a committed relationship for two years. But still, she confesses, "I am emotionally cheating all the time." She scrolls regularly through OkCupid and Bumble. She chats with men. She chats with women. "I have so many friends who think they're in love with me," she notes wistfully, "I don't want to lose them." For Matthew, meanwhile, a thirty-year-old lawyer from England, the infinity of choices available online means there's simply no reason to be monogamous quite yet, no reason to say yes forever to one young woman and no, forever, to all the rest. And because he rarely sees any woman more than once, there's hardly time for his heart to be stolen. Or broken.

Which raises another intriguing aspect of romance in the Internet age. Especially on sites like Tinder, which promise nothing more than a handful of photos and the prospect of a brief encounter, the stakes are remarkably low. People can click or not, match or not, have a coffee or a hookup, and then walk away unscathed. In fact, that's exactly the proposition to which everyone's agreed. No commitment. No regrets. No messy aftereffects of falling in and out of love. As a result, users like Matthew are often inclined *not* to select partners who might truly appeal to them. Because they don't want to fall for someone who might walk away after a sole encounter, don't want to be rejected in favor of someone who is sexier, or taller, or richer, or one stop closer on the subway. Better, in those cases, to play it safe, and either select people you don't care about losing, or scroll and scroll for hours—days, months—without choosing anyone at all. As one young man recently explained: "I had downloaded Tinder with some vague notion of easy hookups but quickly realized that if I struggled to pick up women in a bar, I was still going to struggle to pick them up through my phone screen." Over one holiday weekend, he swiped through a couple hundred women, and never met any of them.[42] One undergraduate woman similarly confided, "The fact that there could be hundreds, if not thousands, of potential dates in your pocket gives an illusion of possibility. In reality, students just become more isolated in a world of fake interactions and awkward run-ins with old matches. We're not getting out of our comfort zone to meet new people. Why approach someone in person when you can hide behind a Tinder profile?"[43]

In the aggregate, these encounters—or lack thereof—are starting to nose their way into demographic data, suggesting emerging patterns of what might be called sexual inequality. Some online daters are having lots (and lots) of uncommitted, casual sex. Some are selecting more carefully, and moving into committed relationships that generally mirror those of earlier generations. But some are essentially dropping out entirely from the dating and mating markets—either because they fail to compete with more attractive online prospects, or because they're scared of failing, or both. In the United States, the youngest members of the millennial generation—those born in the 1990s—are less sexually active than any generation since those that came of age in the 1960s. They are having sex later, less frequently, and with fewer partners.[44] Interestingly, they are also the first generation to have grown up with smartphones.

In Japan, meanwhile, "celibacy syndrome"—*sekkusu shinai shokogun*—has reached nearly epidemic proportions: according to recent surveys, fully 46 percent of young Japanese women and 25 percent of young men are either "not interested in or despise sexual contact."[45] Rather than wade through the complications of dating (and marriage and parenthood), young Japanese are resorting to platonic or insular pursuits: to online porn, virtual reality, or hobbies such as knitting. "I find some of my female friends attractive," explains Satoru Kishino, a thirty-one-year-old male, "but I've learned to live without sex. Emotional entanglements are too complicated. I can't be bothered."[46] Instead, Kishino cooks and cycles, common pastimes for a group Japan's media has dubbed *soshoku danshi*—literally, "grass-eating men."[47]

Elsewhere and more ominously, clusters of men who feel rejected by the contemporary dating market have begun to identify as incels— "involuntary celibates"—and to demand redress for their lack of sexual activity. Arguing, generally, that all men are owed sex, incels have focused their rage at sexually active women, sexually successful men, and the technologies that bring them together. In their most extreme manifestation, several self-styled incels have committed high-profile mass murders, framing their crimes as revenge for their rejection. "All I ever wanted was to fit in and live a happy life," wrote Elliot Rodger in a chilling manifesto, "but I was cast out and rejected, forced to endure an existence of loneliness and insignificance, all because the females of the human species were incapable of seeing the value in me."[48] Immediately after posting these words, Rodger drove to a sorority house near the University of California at Santa Barbara and shot two young women to death.*

For now at least, such violent behavior remains largely confined to society's fringes. But, as is so often the case, behavior on the fringes carries information about the mainstream as well. And the scary truth is that sex may increasingly be skewed along social and economic lines.

Put in historical perspective, this is less revolutionary than it might seem. Because sex, of course, has always been skewed. Men have almost always had substantially greater power in sexual relations than women. White men and women have had power over people of color. Straight people had freedoms that were unknown to homosexuals. The rich have long had favors and access that were denied the poor.[49] But because sex occurred largely within the confines of marriage, and marriage was spread so broadly across society, most people—rich and poor, kings and laborers, men and women—had basic access to sex. It may not have been good sex. It may not have been based in love or even lust. Yet it was still sex, and still spread fairly evenly across any given community.

---

* He also stabbed and killed three men in his apartment before posting his manifesto, shot another woman on the campus, and shot fifteen people during a drive-by spree, killing one additional man. He then crashed his car and shot himself in the head.

By widening these communities, though, and diffusing matchmaking among the matches themselves, online dating has actually—oddly—redistributed access to sex. On the one hand, apps like Tinder and Bumble are incredible equalizers, enabling anyone, with a swipe of their finger, to "like" and be in contact with the person of their dreams. On the other hand, because the person on the other end of the profile has full autonomy to like back (or not), the apps inherently, inevitably, distribute their favors unevenly: some guys get all the swipes. Which is exactly what the data show: women tend to be highly selective on Tinder, generally liking fewer than 10 percent of the profiles they encounter. Men swipe right more often and casually—and match, as a result, far less frequently.[50] The disparity is even more marked in gay communities, where, according to one recent analysis, 2 percent of online daters are having about 28 percent of the total reported sex.[51] Even worse, because apps like Tinder are so transparent, because the platform is essentially built on accumulating points and keeping score, those who don't get the wins—the dates, the mates, the winks and likes and hookup sex—get a particularly painful reminder of where they stand in the social pecking order. As one young Australian man put it more plaintively: "The 10% of highly attractive people fucking all the time make the rest of us feel bad."[52] Age-old sentiments, again. But those 10 percent were less visible in the pre-app era, and the other 90 percent weren't being constantly bombarded by both opportunity and rejection.

Technology alone didn't cause these shifts. Indeed, most commenters (including the most extreme of the incels) are prone, instead, to blame feminism. Or liberals. Or the cadre of neoliberal feminist supporters who supposedly dismantled the conservative values that prevailed in an earlier era.[53] And perhaps, to some extent, they did. But the role of technology here is key. Because it has been technology over the past five decades—the technology that evolved from the Happy Families Planning Service to Grindr and Tinder and JSwipe and more—that has led the way in reshaping patterns of desire and sex. Maybe people like Mark were always covertly interested in polyamory. Maybe Camilla was destined to lust after women as well as men. And maybe those 10 percent of highly attractive people have always been the ones that the other 90 percent wanted to be—*and* to fuck. But all these desires remained unspoken and un-acted-on until apps like Tinder brought them to the fore, showering millions of people with sexual options and preferences they didn't know they had. And now that this genie has escaped from the smartphone, there's no putting it back.

Sinking Noah's Archetype

Go to a restaurant these days—pretty much any restaurant, anywhere in the world. Look at the couples clustered into intimate booths, or the groups clinging around the bar. Now count how many of them are on their phones, either checking furtively each time they get a buzz on their hips or simply scrolling, constantly, apparently oblivious to the actual human beings seated across from them.

It used to be that taking a phone call during dinner was reserved for doctors or expectant dads. Now every trill from Instagram or OkCupid demands an instant response. Or not even a response per se: a nod, a check, an acknowledgment, another mini dose of dopamine. *He likes me!* In the time of Tinder it has become commonplace for two people to agree to meet and then check their accounts during the drink or dinner or hookup, scrolling for a better match. The search for Mr. or Ms. Right has become, in the words of a very old joke, the technologically enabled pursuit of Mr. Right Now. "Termination on demand—instantaneous, without mess, no counting losses or regrets—is the major advantage of internet dating," writes one sociologist.[54]

And so what? We from the Age of In-Person Dating should be wary of passing judgment, careful not to cluck too loudly at behaviors that were never ours to scorn. But the broader question remains: What happens to individuals and society as dating and mating move online? Will the future of romance be substantively different from the past?

Yes, it appears. It will.

The changes will be incremental. They will be peaceful for the most part. As Helen Fisher predicts, they won't disturb the most ancient parts of the human brain, the parts that yearn to love and lust. As people increasingly find each other online, however, two things of great consequence are likely to occur, rippling across patterns of romantic interaction and shifting some of society's most eternal-seeming structures. The first relates to choice, the second to preferences.

What sites like Tinder do most powerfully is to amplify the options for any would-be dater. At any moment, for any user, the sites provide a parade of choices: Bimmerboy and Handyman123, Spanky and Lover-Claudio. Even if some prospects are cringeworthy, even if the appealing ones remain forever out of reach, they are just there. Offering themselves, or some imaginary version of themselves, on and on and on. There was nothing like this—is nothing like this—in real life. That's why people in the physical world of restaurants and bars keep looking at their phones, pulled by the sheer force of possibility. What if the new guy on Hinge is better looking than the one across from me now? What if there's a new message on Bumble? "You're always sort of prowling," confesses Alex, an investment banker in his twenties. "You could talk to two or three girls at a bar and pick the best one, or you can swipe a couple hundred people a day—the sample size is so much larger."[55]

To some people, we know, all this choice can be confusing, even off-putting. At the margins, it can drive those like Japan's *soshoku danshi* or my friend Phoebe into a devoted state of no-dating-at-all. But to others, like Alex, it can also be addictive, fed by the same pings of dopamine that tether users to their Instagram and Snapchat feeds.[56] "It's instant gratification," confesses Jason, a twenty-six-year-old photographer. "You see some pretty girl and you swipe and it's, like, Oh, she thinks you're attractive, too, so it's really addicting, and you just find yourself mindlessly doing it."[57]

Dopamine is a funny chemical. Produced by several tiny regions in the brain, it serves as a neurotransmitter, relaying signals and activating neurons across the brain's cells. When dopamine is released, the brain senses a reward or something pleasurable.[58] Brains in love show highly elevated levels of dopamine on MRI scans; so do the brains of people on cocaine or heroin. Indeed, research increasingly shows that all forms of addiction are effectively dopamine addiction: once the brain grows accustomed to the pleasurable spikes that dopamine produces, it needs those spikes to continue, again and again and again. In the normal course of love, or at least the historical course of love until recently, the dopamine of early attraction—the Romeo-and-Juliet-climb-up-a-balcony kind of love—was replaced and supplemented over time by other, complementary brain chemicals. By oxytocin, primarily, the so-called contentment chemical that floods women's brains after childbirth. And serotonin, a neurotransmitter that lowers arousal and prods individuals to focus intensely on a given partner.[59] But brains on Tinder, like brains on Twitter or Candy Crush, are brains on dopamine. They are brains that quickly grow accustomed to scanning and clicking, hungering for those little jolts of satisfaction. *He likes me. I won. Something good is about to happen and the expectation is giving me pleasure.* Most critically, they are brains that come to crave surprises—breaks in a pattern that prompt the neurotransmitters to fire anew.[60]

We don't know, exactly, what happens to those brains when the dopamine of online dating gives way to more settled affairs. Judging from other areas of addiction, though, one might reasonably expect the wandering brain to want to wander some more, to reenergize itself with a dose of the unexpected, the new. Nearly all monogamous relationships have been vulnerable to the lure of the other; why else would the Ten Commandments include such strict instructions against coveting thy neighbor's wife? But men (and certainly women) in Moses's time hadn't grown up with scores of potential bedmates parading across their palms every day. Their brains and bodies weren't used to responding to a constant cascade of possibility. Those who learn to love on Tinder, by contrast, are deeply and subconsciously accustomed to choice. How will they fare once the demands of a committed relationship take their dopamine away? Or, as Moira Weigel, a millennial author, wonders plaintively: "Can courtship function like a game of musical beds, in which someone turns off the music on your thirtieth birthday, and you have to marry whomever you are lying next to? Does intimacy itself not require any practice?"[61] No, one wants to answer, and yes. Courtship won't work—or at least won't work as it customarily has—if you suddenly ask people addicted to choice to select forever for one. And intimacy does require practice, the hard work of being in love even after the dopamine's gone.

To be sure, many Tinder users are explicitly deploying the app to find long-term, even lifetime, partners—preferences that are even stronger on more traditional sites like eHarmony and Match.com. Many users—men and women, gay and straight—are incontrovertibly settling down, consciously stepping away from the swipe and embracing monogamy. In 2013, more than one-third of all newly married couples in the United States had met each other online.[62] Still, early evidence suggests that a generation raised on conspicuous choice—even if it was only the illusion or fiction of choice—will have a tough time restricting themselves to a world of one and only. "You could say online dating allows people to get into relationships, learn things, and ultimately make a better selection," says Dr. Gian Gonzaga, a psychologist formerly at eHarmony. "But you could also easily see a world in which online dating leads to people leaving relationships the moment they're not working, an overall weakening of commitment."[63] Brian, a twenty-five-year-old teacher, is blunter: "When it's so easy, when it's so available to you . . . and you can meet somebody and fuck them in 20 minutes, it's very hard to contain yourself."[64]

Meanwhile, by allowing participants to fine-tune their preferences so acutely, online dating has also begun to rebalance the distribution of sexual activity. Once upon a time, most people of childbearing age had regular access to a sexual partner—access that was generally arranged through and legitimized by marriage. Men had access to extramarital sexual relations as well, particularly if they were wealthy, while women's options (either before marriage or outside it) were scarcer and more dangerous. In the 1960s, and propelled once again by technology, the sexual revolution exploded this equilibrium; freed from the constant risk of conception, both men and women could have sex more liberally, with a greater number of partners, and without the sanction of marriage. And so they did. By 1975, more than a third of American women reported having had premarital sexual experiences with between two and five sexual partners.[65] Most of this premarital sex, however, was still occurring in a fairly balanced way, and most of its adherents eventually embraced marriage at rates not unlike those of their parents.[66]

Today, by contrast, both sex and marriage are becoming substantially more skewed. Wealthier people are marrying at considerably higher rates than poor ones, beautiful people (or those ranked as beautiful) are having more sex, and larger chunks of the population are being left out.[67] Apps like Tinder didn't cause these shifts, but they have almost certainly contributed to them. Because once attraction is literally a game that all can play, once you leave the confines of the village for the near-infinity of the open market, then the sheer size of this market will tend to channel rewards unevenly, driving dates—or mates or hookup sex—to the most attractive entrants. The handsomest. The richest. The ones most willing to play. It's like an elaborate, extended version of choosing team members on the playground: the good players are grabbed first, then the pretty and popular, then the hangers-on, and then the pathetic few left looking miserably at their shoelaces. In the heterosexual online dating market, preferences have already become predictable and well-documented: men like women who are thin or "toned," and younger than they are. Women like men who are older and more financially secure.[68] Highly educated people generally prefer each other, and no one likes overweight women or short men.[69] After having spent two days on Tinder, one journalist, a self-proclaimed "good-looking guy," wrote sadly, "I got no responses from women. Not one . . . You put a picture of yourself up . . . and nobody finds you attractive. The world decided you're ugly."[70]

Interestingly, Tinder works differentially across the heterosexual gender divide: because men "like" much more frequently than women, men have a disproportionately harder time getting matches, or meets. But because women are "liked" so indiscriminately—it is not unusual for a woman's profile to receive two hundred likes an hour—they have a much harder time finding men who are interested in hanging around after that first date is complete.[71] To be blunter, Tinder has made it easier for women to find sex, but not necessarily relationships. And it's split the pool of men into those, like handsome investment banker Alex, who can get whatever they want, and those, like the Tinder-ing journalist, who sit and scroll and wait. Since 2008, *The Economist* reports, there has been an almost threefold rise in the share of men under thirty who report having no sex.[72]

It is possible that as the Tinder generation ages, they will fall, like their parents and grandparents, into traditional patterns of marriage and monogamy. But I don't think so. They are already, en masse, delaying (or rejecting) marriage: as of 2014, only 26 percent of millennials in the United States were married, compared with 48 percent of baby boomers at the same age, and 65 percent of the Silent Generation.[73] They are having children much later, or not at all—in the United States, per-woman fertility rates fell in 2017 to 1.87; in Japan, they are already hovering at 1.43, well below the rate of population replacement.[74] And they are experimenting with structures and sexualities that were never before possible: gay parents, trans couples, polyamorous relations. The family as we once knew it is liable to be transformed in another generation—or confined, at least, to a much smaller segment, one social configuration out of many. This is the world that Tinder has created. Not by itself, of course, and not in a vacuum. By widening pools of potential mates, however, and changing the ways in which men and women express their attraction, Tinder and its peers have made it hard to imagine going back to a world of slow matching, a world in which every date quietly unfolded against Noah's archetype of two-by-two pairs. In the future, it appears, the social structures that have long surrounded mating and marriage will continue to evolve, possibly to the point of extinction. Our attractions will be shorter and less monogamous, our antennae more primed for the novel and the exciting. Our sexual favors, like other assets, will be distributed less evenly, with the wealthy and good-looking enjoying an ever-greater piece of the pie.[75]

But what, then, of love? Of lust? Of the pings and pangs of falling head over heels for that other soul across the room?

Beautifully, remarkably, it doesn't have to change. As choice-driven apps come to dominate dating, they will undoubtedly prod the objects of our most basic affection to multiply and evolve. We are likely to love more expansively over the course of a lifetime, to fall in and out and in again more quickly and impetuously. Our curiosity and appetites may be whetted more often; the duration of any particular desire may shrink. But love itself—like hunger, like thirst—shows no sign of decline. Instead, it is the structures that have historically defined and channeled that love—courtship, marriage, the two-by-two pairing of heterosexual till-death-do-us-part couples—that are being exploded by technologies such as Tinder and Hinge. Their destruction, or diminution, at least, will be a radical change for the societies that have long been strung upon them. But love itself will settle again, molding and adapting into different configurations that will quickly feel as ancient as desire itself.

Writing in 1955, as modernity was first rocking the norms of love and marriage, Herbert Marcuse predicted that technological evolution would eventually free men and women from the confines of monogamy. Once society was able to produce the necessities of life without the toil of hard labor, he surmised, men and women would at last be able to renounce the repression of "procreative sexuality" and instead explore the boundaries of a new "erotic reality." Fantasies that were once considered perverse, he asserted, would again become natural; the human body would return to its preindustrial purpose, as an instrument of pleasure rather than labor.[76]

Marcuse's utopia, if that's what it was, hasn't quite happened. Perhaps it never will. But the advent of technology has given human beings both more time for leisure and more options for romance. More time to scroll online and wink at attractive strangers. More options to find those strangers, and like them, and maybe even fall in love. As these technologies morph and evolve, they are almost certain to change the ways in which love and sex are channeled and obtained. Love will be less formal in the Tinder-enabled world, monogamy less frequent and revered. Sex will float more casually between people, and distribute its pleasures less evenly. We will love most likely as we always have, as we seem biologically hardwired to do. We will lust. But the structures that have contained these loves and lusts for so long—the forever marriage at the end of courtship's rainbow—is likely to wane over time, a victim of technology's arc as much as the newspaper personal or movie-theater date.

And that's okay. Because the species that was figuratively perpetuated by Noah's Ark—pairing off in preordained configurations of one-man-one-woman-with-a-lifetime-bond-between-them—will learn to reproduce itself through readapted social structures. We will still go forth and multiply, propagating and raising our young to have sex and fall in love and start all over again. How we live as lovers, though, and how we frame our sexual selves will be different—not necessarily better or worse, but altered yet again by technology's wave. By silly apps that sit in our palms and reshape the contours of love.

6

Mad Men

or, How Smart Machines
Are Remaking Masculinity

We are being afflicted with a new disease of which some
readers may not yet have heard the name, but of which they will
hear a great deal in the years to come—namely, *technological
unemployment*. This means unemployment due to our discovery of
means of economizing the use of labor outrunning the pace
at which we can find new uses for labor.

—John Maynard Keynes, "Economic Possibilities for Our
    Grandchildren," 1930

I got a job working construction for the Johnstown Company
But lately there ain't been much work on account of the economy
Now all them things that seemed so important
Well mister they vanished right into the air
Now I just act like I don't remember
Mary acts like she don't care.

—Bruce Springsteen, "The River," 1980

John Maynard Keynes is not generally known for his technological predictions. Instead, the great British economist is remembered for having
conceived of the fiscal policies that helped drag England and the United
States out of the Great Depression and for warning the victors of World
War I against punishing the losers too harshly.[1]

In 1930, however, Keynes also wrote an intriguing short piece on
technology. More precisely, he wrote an essay that aimed to link his

country's economic woes to its technological prowess. Britain, he argued, was suffering the pains of the Great Depression largely because it had developed so successfully. By mastering the great inventions of the eighteenth and nineteenth centuries—"coal, steam, electricity, petrol, steel, rubber, cotton, the chemical industries, automatic machinery and the methods of mass production"—the country had boosted its productivity beyond what its people could absorb; it was replacing jobs more swiftly than it could create them. But the British population would shortly catch up to its newfound riches, Keynes assured his readers, and enjoy the substantially higher standards of living that technology could produce. "We are suffering," he wrote, "not from the rheumatics of old age, but from the growing-pains of over-rapid changes, from the painfulness of readjustment between one economic period and another."[2] Over the long run, he promised, prosperity would return.

Over the long, *long* run, however, the great economist was decidedly less optimistic.* Peering into the future of one hundred years hence—to 2030, a moment that is coming swiftly now—Keynes foresaw that the relentless march of technology would eventually render labor moot. As tools became better and more efficient, as capital intensity grew and productivity spiraled upward, humans would eventually be able to solve the central economic problem of their existence—the struggle for subsistence—without having to toil for themselves. And at that point, Keynes dryly reasoned, "mankind will be deprived of its traditional purpose."

Keynes never speculated about the specific inventions that would bring this future to be. And even with the advantage of hindsight, it's difficult to pinpoint precisely which technologies have brought us to the labor-shifting moment he foresaw. Maybe it was the mainframe computer. Or the semiconductor chip. Or ever-smaller and more powerful chips crammed into ever-more-capable machine tools. The point is that he was right: during the latter decades of the twentieth century, machines

---

* Famously, Keynes also quipped that in the long run we're all dead.

Unimate robotic arms welding chassis on a car production line

that had once relied on human hands began to be run, instead, by other machines. For decades, for example, human workers had physically labored to weld a car's body to its underlying steel chassis. Then, in 1961, a robotic arm named the Unimate #001 was installed on the production line of a General Motors plant, taking over the welding work from its human predecessors.[3] For over a century, similarly, human controllers had been responsible for watching the machines on their factory floors, checking for accuracy and scanning for flaws. Then, in 1968, programmable logic controllers entered the scene and began to perform many of these crucial control functions.[4]

Between roughly 1970 and the present, then, across industry after industry, the advent of information technologies has essentially meant the death of labor. Building, measuring, accounting, surveilling—scores of tasks that once sat squarely in the human realm can now be done as well and more efficiently by "smart" (that is, computer-augmented) machines. Some refer to this change as the Second Industrial Revolution, or the Third—as with waves of feminism, the accounting can get complicated. What is wholly clear, however, is that the industrial world shifted fundamentally between the twentieth and twenty-first centuries, between an era marked by factory labor and one dominated instead by a combination of human programmers and ever-smarter machines.[5]

As these shifts wash across the global economy, the nature and purpose of work will fundamentally change. Far fewer hands will be needed to man the world's machinery, as Keynes predicted; more brains will be required (at least for a while) to design these machines and concoct new forms of leisure. Much has already been written about how these shifts are likely to unfold and how seismic they will be.

But there is another force at play here, too, one somewhat less remarked upon but no less seismic. And that is the way these changes are settling unevenly across men and women, threatening (or promising) to change relations between them in the process. Put simply, the labor shifts of the postindustrial world are falling more profoundly upon men. It is men who are losing the bulk of what were once working-class jobs; men who are less well equipped, by custom if not nature, to adapt to the service jobs being spawned in their place. It is men who don't yet have an identity to replace the stereotypical one of breadwinner, men who will be pushed more overtly to confront the loss of what was once their purpose and privilege.

Specifically, as factory jobs dwindle and smart machines increasingly displace human labor, millions of men are finding themselves not only without a paycheck but also without the patterns of work and family that have long defined them. Ever since the Industrial Revolution, after all, the dominant structure of family life across the Western world has revolved around the gender divide that the factory system first put into place: a male breadwinner with a wife and kids and a solid forty-hour-a-week job. During the course of the twentieth century, as discussed in chapter 3, the advent of household appliances and reliable contraception allowed women to begin to break from this mold and to enter the workforce in ever larger numbers. Changes in reproductive and sexual norms have further transformed both the narrative of the "normal" family and the patterns of its daily life. But changes in industrial technology are liable to deal this model its final blow, and to fall with particular force on the way in which men still define themselves and their roles.

For centuries now, men have been marked primarily by their jobs. Not by their physical prowess or family's title, as was the rule in other times and places. Not by their beauty (as is still so often the fate for women), or by their brains. No, men's status has come from their jobs, and from their explicit role as financial providers. What do men *do*, archetypically speaking, in the industrialized world? They work. They leave the house, go to jobs, and support their families. As Jared Sexton writes in a recent and poignant memoir, "Faced with a job that paid just enough to keep a family afloat, and sometimes not even that, the American man adapted his idea of self-worth to depend on his identity as a laborer . . . Callused hands and tired bones became indicators of self-worth and proof of an attempt, however futile, to do the necessary work to survive."[6] And yet now, as the once-industrialized world shifts increasingly to an information-based economy, this model of man-as-worker is breaking down, destroyed just as soundly as the typewriter or the rotary telephone by technology's advance.

In theory, men pushed out of the manufacturing labor force should be rebelling against the firms that once employed them, the bosses who profited from their work, and the political leaders who refuse to set things right. And to some extent they are. Since the financial crisis of 2008, from both the left and the right, waves of protest have crashed across the industrial world, led, in many cases, by once-working-class men who got left behind—the ones whose jobs and identities have been stealthily replaced by robotic arms, smart machines, and solar power. In time, some of these men could well ignite an actual revolution.

As these forces play out, however, another revolution is also underway, one that is quieter and more personal. Its struggles are less violent, its battles more ideological but no less profound. And its time, some argue, is already long overdue.

This is the revolution to reimagine men, or at least to reconceive a more varied set of possible masculinities. For centuries now, men have been stuck in a very narrow set of norms, one that essentially arose out of the Industrial Revolution and has lingered ever since. Even though these norms never actually applied to all men—even, in fact, when they applied only to a shrinking pool of (usually white, unquestionably straight) working-for-a-living men—they nevertheless defined the cultural stereotype of what a man should be and do. Ironically, perhaps, because women's lives have changed more obviously than men's over the past few decades, women have been able, or forced, to create a range of acceptable personas: the stay-at-home mom, the independent career woman, the proud lesbian. For men, however, the change has been slower. Gay men, for sure, have led the way in carving out new modes of being male. So have men of color who balked at the implicit racial overtones of the workingman mold. But these changes have not yet been sufficient to topple the dominant model, or to force a reckoning among wider groups of men.

Technology, though, is about to accelerate this reimagination. As machines become ever smarter and automation more widespread, more and more men are going to be pushed out of the labor force and into an economy that no longer fits their skills and identities as well as it once did. Some men will be just fine, fashioning the new economy and profiting handsomely from it. Some will be nudged toward different kinds of jobs and different kinds of lives—replacing their breadwinner status, in many cases, with something else. And some will be shoved, sadly and perhaps violently, to the edges of society, victims of a technological trajectory they have no real power to stop.

All these men, though, will also be part of a less visible transformation. Because as they scramble and surf through the postindustrial economy, as they lose jobs and fight back and maybe even assume a considerable chunk of what were once conceived as women's chores, they will be changing the deep and underlying notion of what it means, in the twenty-first century, to be a man.

Invisible Men

As recently as sixty years ago, in the early 1960s, the vast majority of men across the industrialized world participated in the paid labor force, most in the manufacturing sector, and nearly all as heads of their households. By 2010, however, that world had turned topsy-turvy: men's participation in the U.S. labor force had fallen to only 73 percent, and one out of every five men at the prime of what should have been his working life was unemployed—the highest ratio ever recorded.[7] In Great Britain, the share of men ages sixteen to sixty-four who were working fell from 92 percent in 1971 to 76 percent in 2013.[8] In Italy, the unemployment rate for men rose from 5.5 percent in 2008 to 9.5 percent in 2018.[9]

Although elements of this downturn were apparent across nearly all the industrialized world's manufacturing sectors, certain segments, such as auto manufacturing and electronics, were hit particularly hard. These were the industries that faced the double whammy of trade and technology most directly; the ones that had been consolidating production abroad for decades, and whose products—cars, appliances, televisions—were no longer on the cutting edge of consumer desires. Coal mining jobs in the United States, for example, plummeted from a peak of more than 800,000 in the 1920s to around 50,000 by 2017, and iron and steel jobs from 460,000 to fewer than 70,000 over the same period.[10] Even in China, long the world's powerhouse for low-priced labor, firms began moving in the early 2010s to replace humans with less expensive machines. In 2011, for example, Foxconn, the company that builds iPhones and other smartphones, announced plans to introduce up to a million industrial robots to its plants. In the process, it expected, 500,000 human jobs would be permanently eliminated.[11]

Not all these jobs, of course, belonged to men. In fact, women had been making serious inroads onto factory floors and assembly lines for decades.[12] But much of manufacturing's core labor force remained male, and it was disproportionately male laborers who bore the brunt of automation. In the United States, fully 82 percent of the jobs lost during the 2008 financial crisis had previously been held by men—largely in the manufacturing and construction sectors.[13]

In theory, any of these men might have wrested themselves back into the workforce. Indeed, ever since the earliest days of industrialization, each successive wave of technological innovation has been followed by shifts and reconfigurations of the labor force, with workers displaced by one round of invention—from steam-powered looms to automobiles—eventually finding new sources of employment. Textile workers in England, for example, who rallied around the probably mythical figure of Ned Ludd in the 1810s to protest the automation of their mills—the first, actual Luddites—moved within a generation or so to other sectors of the country's booming industrial economy.[14] Automatic threshing machines replaced nearly a third of the agricultural labor force in the middle of the nineteenth century without any long-standing effect on overall employment.[15] Even the proverbial buggy whip makers, laid off forever by the advent of automobiles, gradually found work in industries spawned by new modes of transportation: building cars, or conducting trains, or constructing millions of miles of new roads. Employment itself did not decline when whips and buggies went away.

This time, though, the shift is more profound. Because this time, technology is creating products and industries that need far fewer humans to power them: software like TurboTax can replace the work once done by tens of thousands of tax preparers without an equivalent uptick in the number of software designers. Online retailers such as Amazon or

Alibaba can wipe out entire retail sectors and create, in return, only a minimal number of jobs for programmers, managers, and warehouse attendants.[16] Marx or Lenin would see this as the oppression of the proletariat by the forces of monopoly capitalism. Most economists would describe it as a step-change in the relative position of labor versus capital, a simple corollary, as Larry Summers has written, of "output rising and wages falling."[17] For many former workers, though, it means simply that the jobs and the world they once knew are never coming back.

This is the fundamental change being wrought by our postindustrial economy.[18] As automation proceeds, and machines perform more and more of the work that was once done by men, opportunities for employment are splitting into two widely disjointed chunks. On one side are the jobs that involve directing the machines: computer programmers, marketing managers, industrial designers, software engineers. On the other, jobs that, for the moment at least, machines cannot do. Hairdressing, for example. Nursing. Brain surgery. Middle school teaching and pastry making. For many people, the problem is that they cannot shift very easily into any of these alternative paths. The direct-the-machine jobs involve a level of education that, by middle age, is realistically out of their reach. And the don't-compete-with-the-machine jobs are either out of their reach again (brain surgeon) or below their expectations (hairdresser, pastry maker). Which is where the gender piece of this shift becomes so important.

Simply put, women have thus far been markedly more inclined to adapt to the employment challenges of a postindustrial world. Although men undoubtedly still hold a disproportionate share of the jobs at the very top of this economy, women are moving more flexibly across its middle and lower ranks.[19] They are earning more college degrees than men, which means they are better positioned to gain entry-level jobs in growth fields like teaching and nursing.[20] They are more likely to enroll in retraining programs that can open the door to well-paying and stable fields like pharmacy.[21] And they are the bulk of the workforce in so-called helping professions like home health care and administrative assistance—

lower-paid, but largely resistant to competition from machines.[22] As a result, as writers like Hanna Rosin and Liza Mundy have argued, we are moving into an increasingly bifurcated workforce, split by both gender and earning potential. At the top end of the economy are the knowledge workers, those directing the machines or doing the highly skilled jobs that remain, still, beyond automation's reach: brain surgery, physics, championship tennis. Entry into this world is limited by education and privilege, and earnings flow disproportionately to those at the very top.[23] The second economy is composed of what used to be the working class. Construction and manufacturing remain part of this sector, but jobs there are shrinking fast, replaced (insofar as they *are* being replaced) by lower-level service work. By housekeepers and baristas, personal trainers and security guards—good, solid, decent jobs, but not the kind that generally offer regular hours, or generate sufficient income to provide comfortably for a family of four. Not the kind of jobs that allow a man to be breadwinner to his wife and children in the way that generations of men took for granted, and presumed would last forever.

Instead, the men who worked in the factories are slowly settling into new occupations and family patterns. Many of them work fewer and more irregular hours, as handymen or Uber drivers, contractors or consultants. Many split what used to be their workday with a spouse, working alternating shifts to increase the family income. And many, particularly those with fewer skills, remain stubbornly under- or unemployed. As one recent article explained, "The jobs of the future are going to women, while their male counterparts, in seeming contradiction to their own self-interest, cling to remnants of the economy of the past."[24]

Politically, these men are not yet igniting any kind of sustained protest. They are not marching on state capitols to demand that lawmakers create jobs or dismantle the technologies that displaced them. In fact, insofar as they are organizing, many of technology's victims don't even acknowledge the role it has played in their own shifting fortunes.[25] Some are turning to the solace of drugs or fringe political groups. But many are also quietly wedging themselves into the new structures that technology has wrought. And in doing so—in grappling every day with how to work, and support a family, and build a new identity—these post-industrial workers are in fact leading a revolution of a very different sort.

Postcards from the Edge

For more than twenty years, Tim Saunders was your average working dad. He spent his days at the bustling Maytag factory in Newton, Iowa, shaping sheet metal into doors and cabinets for the dishwashers that had once made Newton the "Washing Machine Capital of the World."[26] He earned a solid wage of twenty-three dollars an hour, enough to support his wife, Rhonda, who worked only part-time, and their two children. The family was never rich, but they enjoyed camping trips and cookouts, and occasionally splurged on big-ticket purchases, like a used racing car for their son and a trip to New York for their daughter. Both kids were heading to college, relying on a combination of student loans and Tim's hard-earned savings.[27]

In 2006, however, the Whirlpool Corporation bought Maytag and began to consolidate its operations. Whirlpool's management reportedly "looked at every element of the business." They "did the evaluations [they] needed to do," according to a company spokesperson, and calculated a strategy for producing "best cost, best quality, best distribution."[28] And that formula, apparently, did not include the Newton plant. Between December 2006 and August 2007, the company laid off twelve hundred workers; on October 26, 2007, it closed the plant for good, consolidating its operations into similar plants in Mexico and Ohio and dismantling a workforce that had once included over three thousand people. "Life was good," recalled David Daehler, another longtime employee. "We was working 24/7 at the first plant I worked in. And we couldn't put on enough product . . . [But] when things started going down, they really spiraled."[29]

Within months, as their savings dwindled and severance ran out, the Saunderses and Daehlers were forced to adjust. Rhonda Saunders took on two jobs, earning twelve dollars an hour in the accounts payable department of one small manufacturer and keeping the books for another, while Tim enrolled in computer programming classes at the local community college and tried to help out more at home. Lori Daehler took on jobs as a housekeeper, while David looked for work out of town.

Their stories, of course, are only that: stories of two families struggling to make ends meet in the wake of a single plant's closure. But in the early decades of the twenty-first century, the narrative of their personal lives became increasingly common. In Alexander City, Alabama, over six thousand jobs were eliminated by the gradual shutdown of the Russell Corporation, an athletic-wear manufacturer that was once the region's largest employer.[30] Most of the men affected by the plant's demise remained unemployed a decade later, spending their days idling at home or looking for work. Most of their wives, by contrast, jumped into the workforce, finding jobs as teachers, or secretaries, or social workers.[31] In former industrial nations like Latvia and Estonia, rates of unemployment among men climbed into the double digits in 2008 and 2009, while female unemployment climbed only slightly.[32] Even in Germany, long the most stable industrial producer in the world, the percentage of the labor force employed in the industrial sector slipped from 23 percent in 1980 to just 15 percent by 2012.[33] Nearly all of the fastest-growing jobs in the United States are now projected to be in fields—occupational therapy, home health care, nursing—that have typically been dominated by women.[34]

Despite how they might define themselves, therefore, Tim Saunders and David Daehler and millions like them *are* forging a revolution, even if it's one they never imagined having or wanting to initiate. Because, reluctantly or not, they are changing some of the oldest and most enduring structures in society: how families are constituted and gender roles assigned. Who makes the lunches? Dries the tears? Pays the bills? Within the confines of any given family, these choices feel simultaneously trivial and profound—huge for the individual who is suddenly mastering peanut butter and jelly constructions, but insignificant beyond his kitchen. In the aggregate, though, these decisions connect. And as technology's arc shoves more and more fathers into reconfigured roles, the larger structure of the family—how roles are conceived and genders defined and power displayed—is bound to shift as well.

Of Shifting Gears and Changing Roles

If you walk through the hallways of just about any major university these days, you are bound to encounter a Department of Gender Studies. Or Gender and Sexuality Studies. Or something with a similar name that didn't exist as recently as forty years ago.[35] These are the creations of a generation of feminist scholars who entered the cultural mainstream in the 1970s and began to specialize in women's issues: women's history (sometimes called "herstory" by the more radically minded), women's literature, women in media or politics.[36] Over time, these disciplines grew larger and more robust, tackling female-focused topics that ranged across an interdisciplinary spectrum and generating rich bodies of both research and theory.

Outside academia, meanwhile, feminism has also slipped widely into popular discourse—even if couched under less politically loaded labels like "women's leadership" or "girl power." Today, across most of the world, there are powerful advocates for what is (sometimes derisively) called "corporate feminism"—focusing mostly on the plight of well-educated women in the corporate workforce—and adamant, persistent cries for gender pay equity.[37] There are high-profile campaigns for reproductive rights and mainstream energies devoted to increasing women's presence on corporate boards and in the media. In 2017, the #MeToo movement struck what could be a turning-point chord across Hollywood and beyond, forcing men to reckon at long last with the consequences of workplace sexual assault and harassment. Which isn't to say that the fight for women's rights is either over or won. On the contrary, as countless studies have noted, women remain stubbornly stuck outside the corridors of power, and shoulder a disproportionate share of their families' child care and housework. They are still struggling to have it all, be it all, and to juggle, as I have written elsewhere, the "ever-escalating expectations that society thrusts upon them."[38] But at least we now understand the challenges and constraints that women face. At least we recognize, both individually and as societies, that women's roles are changing, and that change is hard. We have put a name to Betty Friedan's "problem that has no name."[39] And we have started—slowly, painfully, in frustrating stops and starts—to solve it.

By contrast, we know comparatively little about the tensions inherent in our dominant norms of masculinity, and about how men's roles and identities are evolving in the wake of both societal and technological change. Perhaps, as sociologist Victor Seidler has suggested, this is because men's power has been accepted and institutionalized for so long that we haven't carved space to examine the broader range of male experiences and masculinities.[40] Or, as R. W. Connell surmises, because men inherently realize that analyzing their power would likely entail ceding some of it to others.[41] But for whatever reason, although there has been a noticeable uptick in the past decade or so, the field of what is variously called men's studies or masculinities remains relatively anemic, and focused largely on men who fall beyond the borders of what was long seen as traditional masculinity. So there are studies of gay men, trans men, immigrant men, but perilously little, by contrast, on changing norms of masculinity, or on the plights of men pushed out of the workforce.[42]

At a gut level, we recognize the archetype of this Traditional Man, the man who defines the cultural trope of industrial masculinity. He is white (almost always), straight, married, and working for a living. He looks a lot like Tim and David, and the nameless hordes who populate every beer and car advertisement you've ever seen. We know Traditional Man because we have lived so long in a world defined by his presence and carefree privilege: even now, white men in the United States get paid, on average, 28 percent more than white women, and 37 percent more than Hispanic men.[43] But we know this man as an archetype, a cultural trope—not as a flesh-and-blood human who is trying to figure out how to pay the bills and spend his days once his job at the factory is over. In both academic discourse and more popular accounts, we haven't fully wrestled yet with how our traditionally narrow view of masculinity is liable to evolve once it has been buffeted by the joint forces of deindustrialization and societal change, or how the men caught by these changes are liable to respond.[44] Even worse, much of the discussion that does exist around men and masculinity has recently taken a sharp and horrifying turn toward the alt-right.[45]

How is this possible?

Part of the problem, I suspect, is that the link between technology and the changing workforce has not yet been strongly and explicitly established. So when men lose their jobs, or face a fundamentally more challenging labor market, they take it personally and retreat—either into the relative anonymity of their own homes, or into one of the few outlets available to them. They rally, if they rally at all, around vestigial identities—the fighter, the patriot, the hunter—rather than exploring new ways of imagining themselves and their roles. As Jared Sexton, a millennial writer raised in a hardscrabble working-class family, recalls: "Men were effectively the kings of their household, the final word on every matter, above reproach or question. They were to be feared and taken care of. They sat in front of a game on the TV like a royal contemplating stately matters upon a throne . . . Like kings, they were benefactors of privilege they had earned by simply being born. And when those privileges were tested . . . they declared war."[46] Finding a way to rechannel and redefine this deeply seated identity has thus far proven difficult.

Women, by contrast, have had several decades already to find ways of reconfiguring themselves and their roles. If we trace feminism's rise at least in part to the advent of household appliances and secure contraception, then what propelled women into reimagining their lives was the work of early feminists like Betty Friedan, who named the problem she saw around her and suggested ways of responding. In particular, Friedan realized the extent to which household technologies had shrunk women's work, leaving them with time and energy to spare. She realized that advertisers—using the then-innovative technology of broadcast television—were selling women an illusion of household bliss, urging them to spend more time waxing their floors and polishing their silver when in fact they could now spend less. And she urged women—passionately, poignantly—to create a different identity for themselves, exchanging the constraints of kitchen and laundry for "truly challenging work."[47] In other words, she realized that technology had created the makings of a new world, and she called upon women to seize it. Every subsequent feminist argument, even those voiced by writers who despise Friedan, starts from this basic premise: the world has changed, giving women, at last, the means to re-create the purpose and identity of their lives.

Men haven't changed yet because they haven't had to. The technologies that altered women's lives in the 1950s and '60s didn't touch men as much; because they weren't doing housework, they couldn't really tell that the hours necessary to do a family's laundry or bake its bread had plummeted. Because they had never borne the risks of pregnancy, they weren't freed to the same extent by access to the pill. Instead, the technologies whose brunt affected men most directly were those of a more recent vintage. Computers. Machine tools. Information technologies—the ones whose development has helped to render their jobs obsolete. These technologies have only really come into prominence, however, since the mid-1980s. And because their effects are more ambiguous and diffuse, they haven't yet generated the kind of radical soul-searching that feminists like Friedan ignited. And without that—without any alternative vision of what it means to be a working husband and father—husbands and fathers who are no longer working have only had older images to fall back upon, and to mourn.

Hanna Rosin captures this tension poignantly in her book *The End of Men*. She tells the tale of David, a twenty-nine-year-old Canadian who describes himself as "adapting pretty well to the new world order." He works as an editor, and has all the perks of a straight, privileged millennial: a dog, a girlfriend, an apartment, and a car. But his girlfriend, Clare, earns more than he does, and he is already brooding about a possible future in which she stays in the workforce and he looks after the kids. David couldn't care less, Rosin reports, about things like "head of the family" or "patriarchal authority." But still, when he passes an apparently happy dad at the playground, he cringes. "I'm progressive and enlightened, and on an ideological political level I believe in that guy," he confides to Rosin. "I want that guy to exist. I just don't want to *be* that guy."[48]

Recently, a handful of men have started to examine what it might mean to be that guy or, more usefully, how new models of manhood are emerging and being defined.[49] Writing in *The New Republic*, for instance, Marc Tracy notes somewhat wistfully, "Most men stress over the next step in their professions, with the attitude that if they happen to fall in love and settle down, well, that's great, too. But recently . . . we have begun to question whether our most basic priorities aren't out of whack, and to wonder whether, for reasons both social and surprisingly biological, we shouldn't be as 'ambitious' to have children as we are to land the next great job . . . Do we want it 'all'?"[50] Similarly, in describing why he decided to collect a series of essays on the "good man," James Houghton recounts that "it seemed that the men of our generation spend a lot of time struggling to balance the competing interests of achieving professional success and being good husbands and fathers and sons. And unlike women, who are much better socialized to talk about how these same pressures affect them, we tend to keep those burdens to ourselves."[51]

But what of men for whom the ideal of success is fading fast? For the Tims and Davids whose manufacturing jobs have almost certainly now been lost for good? These are the men who long hewed to the cultural tropes of masculinity, the men, as Michael Kimmel notes in *Angry White Men*, whose "dream became a nightmare of downsizing, job loss, outsourcing, plant closing . . . losing the family farm." These are the men, as Bruce Springsteen sings, who once believed in a "promised land," and whose rates of suicide, opioid addiction, and depression are instead now soaring. Between 1999 and 2013, the death rate for white American men between the ages of forty-five and fifty-four actually rose, mostly as a result of drugs, alcohol, suicide, and liver disease, with the greatest increase coming from those with the least education.[52]

As a society, we don't yet have good options—in terms of either job opportunities or cultural archetypes—for what the former working-man might be. And until we do, both he and the rest of us will be ensnared by the struggle to reclaim a past that is never coming back.

In many ways, then, the proverbial circle has started to come around. Women have now spent over fifty years thrashing through the problem that once "had no name"—arguing and debating and ultimately carving new and far more flexible models of what it means to be a woman and a mother and a wife. What was so often neglected in the women's conversations, however, was the role of men, and the extent to which changing lives and career patterns for women would inevitably demand that men change, too. That reckoning didn't occur as a direct response to feminism. In fact, in many segments of society, feminism remains a thorn in the side of men, a nasty reminder of the "nasty women" who are seizing what used to be theirs.[53] But technology has now completed what ideology began. And as the jobs that once defined the essence of manhood increasingly disappear, men will need to name the problems they face and begin the job of addressing them.

When the Machines Shut Down

At the end of World War II, the industrialized world was just that: large swaths of the globe characterized by economic systems that relied on industries and factories and machines created by the first Industrial Revolution. It was a man's world, both archetypally and in fact.

By the turn of the twenty-first century, though, that man's world had started to fray. Women were streaming into what had once been all-male bastions, and growing numbers of openly gay men were celebrating and showcasing fundamentally different models of masculinity.[54] Together, these two forces alone would have forced significant change in how society thought about and defined men. But they weren't alone. Instead, the turn of the century also experienced one of history's greatest technological shifts—the turn from industry to information, from machines run by men to those explicitly engineered to replace them. In the digital age, jobs based on brawn will increasingly fall prey to those based on brain, or on information technologies. And millions of muscle-bound jobs are simply disappearing, along with those men who long relied on their muscles to get the job done.

It is too early to predict just how this world will evolve, or how next-generation men will readjust their roles and relationships. Almost certainly, though, technology's effects will be mediated by the market, and by where various men stand in the pecking order of privilege.

For instance, it already seems clear that whatever replaces the stereotypical male icon of the twentieth century will be less fixed and monolithic. Because the economy is shifting so dramatically, uprooting industries and entire sectors, men (along with women) will be re-sorted into a wider range of categories and occupations. Some, for sure, will remain primarily fathers and breadwinners. Some will be fathers but not breadwinners. Others, fathers in terms of biology but not living arrangement. And so forth. As these permutations begin to morph and multiply, however, a variety of masculinities is likely to emerge.

One of these might conveniently be labeled the Man in the Gray Fleece Vest. We know this fellow already, have watched him on CNN and read about his exploits in every business publication. Fleece Man is brash and generally young.[55] He lives in Silicon Valley, or one of its echoes across the world, and works in the technology sector. He is wealthy and well educated, and is the closest living descendant of the Man in the Gray Flannel Suit. But his workplace is different, as are his norms of work. Fleece Man doesn't work nine-to-five, but instead labors around the clock, tethered by technology to whatever issues might arise. He is mobile, both geographically and career-wise, following the pull of deal flows and start-ups, and he intends to retire before the age of fifty.[56] He may be married—to a woman or man—but he doesn't show up frequently for soccer games or school conferences. He's just too busy.[57] Fleece Man is the wealthiest of the new male archetypes, for he is the one blessed, and indeed created, by a world of smart machines.

Outside these rarefied circles, though, men in the information economy are likely to be scrambling to find new means of employment and models of intimate life. Unlike their fathers, these men in the middle will probably not spend their entire careers in a single field, or as their families' primary breadwinners. Instead, they are liable to bounce, as most already are, between jobs and sectors; between bringing home the bacon and frying it up for a quick dinner. They may remain single; they may father children either in or outside of marriage; they may settle into otherwise conventional marriages in which their spouse becomes the primary breadwinner. In many ways, their lives will resemble those of contemporary women more than traditional men. Because they will be the ones juggling now, slipping out of meetings to make it to the orthodontist on time, or declining a promotion to care for an aging parent. They will be the ones shifting identities across the arc of their lives, and acquiring new skills as circumstances demand.

There will be massive variation, of course, across this range of men, driven by factors that include geography, race, and sexuality. There will be gay men and straight men in this category, single, queer, and trans men all trying on and fitting into newly structured roles and responsibilities. The biggest variations within this group, though, will probably cluster around class, and stem from the sharply bifurcated economy of the information age. Some men, not as well situated as the Fleece Vest class but still well educated and well positioned, will negotiate with their partners (typically also well educated and well positioned) to reformulate and rejigger what was once a predictable slate of lifetime tasks. These are the erstwhile lawyers and doctors and editors, the men who live in Brooklyn and Stockholm and Barcelona, and are partnered with lawyers and doctors and editors who also want or need to work. They are the ones gamely piloting strollers around the playground and taking charge of the dental appointments, the ones carving new roles partly because they have been thrust in that direction by technological and economic change, and partly because they have embraced the ideals behind the change. Reports are all anecdotal at this stage, but many men in the middle, particularly of the millennial sort, attest to the joys that juggling can bring. "When I first started my business," writes one recent dad, for example, "I was concerned that these two roles would be in major conflict, but what I have found is that my daughter energizes me to go to work every day and crush it."[58] Like the women whose lives were recharted by feminism, these men are simultaneously overjoyed, frustrated, and baffled by the choices that confront them.[59] In time, they may well be liberated by them, too.

Further down the economic ladder, though, are men whose juggling brings them closer to the edges of financial duress. These are the men like Tim and David, men who have been pushed more aggressively out of the postindustrial economy, and whose embrace of change comes more from necessity than choice. They are taking on what once was women's work—doing laundry, making lunches—because that work remains to be done, even as their jobs have disappeared. Most of these men are significantly worse off in the new economy than they were in the old; their struggle to carve new modes of manhood is about necessity as well as identity. They need to find new jobs—many of which exist only in what were once women's fields—and squeeze their paid working hours into a growing jumble of family responsibilities. These men in the middle need to continue supporting their families, but in less strictly defined ways—ways that allow for part-time financial contributions, or emotional contributions, or just-wait-until-I-get-additional-training contributions. None of which aligns very well with the traditional archetypes of manhood, clustered tightly as they are around the image of man as provider.

Finally, there is a group of men who are plotting, or at least imagining, a reactionary return to the world that's been lost. They are the neo-Luddites of the postindustrial age, the ones left behind by technology's advance and unwilling, or unable, to adapt. We see these men at nationalist rallies, and hear their determined cries to Make America (or England or Greece or Japan) Great Again. In 2008 and 2012, Jasper County, home of Newton's Maytag plant and other once-mighty giants, voted squarely for Barack Obama. In 2016, Trump won the county by a landslide.

Few of these left-behinds link their economic misfortunes to technology's advance. Indeed, they are far more likely to blame immigrants, or women, or politicians, or Wall Street. Rallying his troops before the catastrophic 2017 march in Charlottesville, Virginia, for example, one organizer urged all "nationalists and patriots" to "take a stand . . . for our European identity."[60] Similar cries have been voiced in France, where supporters of the far-right Front National regularly blame their country's purported ills on foreigners and illegal immigration, and in the United Kingdom, where arguments in favor of Brexit have centered largely around immigration or the stifling hand of European bureaucracy.[61] Many of these arguments and their advocates are simply racist, emboldened by the times to voice sentiments that are usually shunned and silenced.

Yet despite what these protesters may say—despite what they may think—much of the blame for their discontent lies with technology's inexorable advance. Because it is technology that has taken their jobs away and thrust them to the lower rungs of the economic ladder. It is technology—information technologies, machine tools, automation, and digitization—that has rendered their skills less valuable and made it more difficult for them to learn new ones. And although technology has ironically helped to amplify their voices and political reach (through social media platforms such as Twitter), technology is almost certainly not going to give them their jobs back—at least not this current disaffected generation.

With their economic prospects so firmly compromised, the left-behinds face an existential dilemma, much like the one that faced their Luddite brethren over two hundred years ago. They can fight the future, grabbing on to slogans and leaders who promise to ferry them back to the past. They can cloak themselves in the icons of an earlier manhood—a still-industrial, or even preindustrial, era that lauded men for their physical strength and their machine-handling prowess. They can deny feminism in all its guises, and revile the women who dare to attain power. But in the end, the men whom technology left behind cannot win that fight. Because the postwar world of manufacturing is gone—gone as clearly and definitively as the once-equally-mighty worlds of hand looms and buggy whips. The left-behinds can protest this change. They can mourn and rally against it. Or they can join the vanguard of the other revolution, the social revolution, and strive to re-create themselves in what is now a new world.

It is eerie to read Keynes today, and to count all the innovations he could not possibly have imagined: Nuclear weapons. Smartphones. Nano-technologies. But he got the basics frighteningly right. We are nearly at the end of the time period he prophesied, and clearly reaching the point at which technology will have assumed much of the physical toil that once was humankind's, leaving the rest of us—and particularly the men—discarding the habits and identities they have borne for so long and searching for new ones.

Men haven't found their postindustrial prophet yet, the male counterpart to feminism's many oracles. They haven't generated a Beauvoir or a Friedan, someone who will name their pain and chart a path forward. And so for now, at least, there may be wisdom enough in the words of Bruce Springsteen, an artist who has captured the decline of the iconic workingman ever since that decline began. "We said we'd walk together baby," he sang in 1992,

> come what may
> That come the twilight should we lose our way
> If as we're walking a hand should slip free
> I'll wait for you
> And should I fall behind
> Wait for me.

Part III

The

Way

We

Will

Live

7

Trans*itions

Several years ago, a young man named William asked to see me. He had missed my regular office hours, and wanted time to talk about an administrative problem. The people in my office were reluctant—he was being rather aggressive on the phone, they feared, and wasn't really supposed to get an appointment that wasn't available to the other twenty-six hundred students on campus. But I was intrigued, even curious. Because I was president of an all-women's college. And William, apparently, was a man.

Certainly, he looked like a man. He came to the office in a leather jacket and Converse sneakers, with a messenger bag slung low across his chest. He had close-cropped curls and a full reddish beard stretched across a distinctly masculine jawline. And his voice, steady and well timbred, was low.

We had a surprisingly nice chat, especially since he had essentially come to ask for the impossible. William, it turned out, had transferred to Barnard only eighteen months earlier, leaving a fine co-educational university to come to our small, all-women's college. He had been Carly at the time—born a girl, raised a girl, and never in doubt of his girl-ness. But then, over the next few months, Carly decided that she was really a he, and that the emotional struggles she/he had been experiencing since puberty were caused not by depression or anxiety or fear of being gay, but instead by gender dysphoria—a condition in which an individual feels strongly that their body and gender do not match. And so, with the financial and emotional support of her/his surprised but accommodating parents, Carly had undergone hormonal and surgical treatment and become—in less than a year—William. His only remaining problem, he reported, was that his college transcript now effectively outed him as a former girl. A trans man. A person who had done what, until recently, could only be imagined as a fantasy or a nightmare. He had changed his sex.

Presumably, the world has always known people like William—people who are born in what they consider to be the wrong kind of bodies, with the wrong sex attached. India, for example, has long recognized the existence of *hijras*, or "third-gender" individuals; Samoans embrace *fa'afafine*, who are born male but embody distinctively female traits. Yet, for millennia, there was nothing that the *hijras* and *fa'afafine*, the shes-who-would-be-hes and hes-who-would-be-shes, could do about their status, or the sadness that so frequently accompanied it. But then, around 2010 or so, science stepped in, beginning in the Netherlands and then sweeping across western Europe and the United States. Doctors started to treat gender dysphoria not only as a medical condition but, crucially, one that could be fixed. Already, a handful of plastic surgeons around the world had quietly been rearranging the most obvious physical manifestations of trans men and trans women, removing or adding breasts (known colloquially as "top surgery") or, less frequently, replacing genitalia (the "bottom"). A few gynecological and urological surgeons were removing ovaries or testicles from the women or men who no longer wanted them. And doctors of various types were prescribing hormones that allowed both trans men and trans women to become, endocrinologically at least, closer to the gender they saw as theirs. But it was doctors in the Netherlands who brought these various practices into a whole, and launched a distinctive protocol—starting, ideally, before their patients had even entered puberty—to give individuals the physical bodies and identities that matched the ones in their minds.

Today, trans men and trans women still face persecution and discrimination; they are restricted in many parts of the world from using the restrooms and other facilities that feel appropriate to them, and suffer from disturbingly high rates of suicide and murder.[1] Men and women who have undergone sex reassignment surgery often find it difficult to find employment in their postsurgical state, and many can't afford the treatments they so desperately desire. Despite these setbacks, however, the early decades of the twenty-first century witnessed a veritable revolution in gender, a transformation not only of the physical bodies of the men and women involved but also in the very definition of men, and women, and those who might fall between them on what was once a clear-cut gender binary.

Unlike other areas of social transformation, the trans revolution arose from a gradually developing body of scientific research and medical experimentation. There was no equivalent to the steam engine in the field of gender, no breakthrough like IVF that hinged on a single, highly visible technology. Instead, the science that eventually enabled girls like Carly to become men like William was part of the broader revelation of biomedicine and biochemistry that unfolded across the twentieth century. Its pioneers were scientists and doctors, some focused on what their research meant for gender and its discontents, and others working to decipher more basic units of analysis: genes, and embryos, and what would eventually become known and understood as hormones. Much of the pathbreaking work was done by German and Austrian scientists in the 1920s and then brought to a horrifying crash by the Nazis' condemnation of any research that didn't advance their own genocidal intent. The science then crept back slowly, breaking into public prominence around the turn of the twenty-first century, when trans individuals like William were suddenly making headlines and challenging accepted practices around the world. In 2014, India's Supreme Court instituted a third gender option on all government documents and Internet giant Facebook replaced its traditional menu of gender options—*male* or *female*, like gender menus most everywhere—with a considerably longer list of fifty possibilities, including transgender, cisgender, pangender, and androgynous.[2]

For many people, the changing contours of gender identity are both complicated and threatening. On the right, reactionaries paint people like William as freaks or perverts—sexual deviants in search of something that isn't right or normal. On the left, progressives are embarrassed to admit that they don't quite understand the sexual part of the gender change, or its links (or lack thereof) to homosexuality.* The words are confusing—*transgender? transsexual? transvestite?*—as is the implied importance of changing things that most people take for granted. Sensitive things like their breasts and their genitals and the ways their bodies perform the most intimate of tasks. Things that no one really wants to talk about in public, or imagine being transformed. But that's exactly what transgender individuals have done, or dream of doing. And their dreams have made the rest of us reconsider our realities.

Because what is truly revolutionary about transgender people is not

---

* As will be discussed later in this chapter, transgender individuals can be either heterosexual or homosexual.

the people themselves, but rather the broader transitions that their lives and actions will inevitably call forth. As the Williams of this world multiply and diverge, they will compel the rest of us—those who once fell easily and wholeheartedly onto one side of the gender divide or the other—to realize that the divide itself isn't as solid as we once believed it to be. We will be brought to understand that the distinction between male and female isn't necessarily carved into an individual's biology or brain, and that people may shift over the course of their lives from one gender identity to another. More radically, as biomedical science continues to evolve and respond to societal demands, we may see gender itself begin to morph and dissolve, replacing the traditional boy-girl binary with something far more fluid and diverse. And at that point, everything we once presumed to be true about men and women, sex and gender, will be up for grabs, shaking not only how we live our lives but also how we conceive of them, and of ourselves.

A Quick Note on Words

In any contemporary discussion of transgender issues, even the most basic words can become both confusing and contentious. This is due partly to the newness of the terms themselves, and partly to the politics that surround them. Many trans advocates, for example, reject the term "transgendered," as either a verb or an adjective, arguing that such usage implies unusual things being done to or experienced by an individual—as in, "Jose was a transgendered male" or "He transgendered after an unhappy youth." Instead, they urge using "trans" as an all-purpose term for those who do not fit the customary gender binary, much as "gay" has become an all-purpose term for both male and female homosexuals.

Within the trans community, pronouns also remain complicated. When Carly became William, he had to shift quickly from female pronouns (she/her/hers) to the male equivalents (he/him/his). But what set of pronouns works during the transition phase, or to refer to the time when the now-William was still Carly? (This is precisely why William had such a problem with his college transcript, since the words revealed his former life as a woman.) What, too, to do with people like Maj, another of my former students, who happily defined him/herself as neither male nor female, and preferred to be referred to as an it? Some languages offer the adaptable convenience of gender-neutral pronouns: in Finnish, for example, "hän" means both "she" and "he." In Turkish, "o" refers equally to "she," "he," and "it." Clunky English has few options, prompting recent demands for new sets of words: ze/hir/hirs, in one combination, or zhe/zher/zhim.

Finally, there is a critical but often misunderstood difference between the terms "gender" and "sex." Historically, "sex" was used to describe both a person's genitals and their identity. Men had penises, displayed masculine characteristics such as aggression and strength, and desired sexual intimacy with women. Women had ovaries and vaginas, displayed feminine characteristics such as sweetness and docility, and desired sexual intimacy with men. As homosexuality became widely acknowledged, the boundaries of sexual discourse changed accordingly, and it became increasingly common—if still not always accepted—for those who were sexually male to desire and have sex with other males. Linguistically, they were simply "homosexual," as were women who preferred to have sex with other women. The emergence of transgender people put pressure on this dichotomy, since their biological sex (penis or vagina and corresponding reproductive organs) did not necessarily align with their sexual orientations. Trans women, in other words, might want to have sex with men—but they didn't identify as gay men and occasionally even had a deep-seated repugnance to their own male genitalia. To clarify this state of affairs, the term "gender identity" was increasingly used to refer to an individual's sense of his or her (or their or its) own sense of being. A person of the male sex, in other words, could now identify as being female, just as a biological female could identify as being male—and gender identity did not necessarily correspond to sexual orientation. As Dr. Norman Spack, who chairs the Gender Management Service at Boston's Children

Hospital, frequently reminds people: "Sexual orientation is who you go to bed *with*. Gender identity is who you go to bed *as*."

For the purpose of grammatical convenience, I will use "transgender" throughout this chapter as an adjective, referring to those individuals who identify with a gender that differs from their biological sex, regardless of whether they have undergone either surgical or hormonal treatment. I will use the term "cross-dresser" rather than the archaic "transvestite" (which is generally now considered to be a slur) and will use the gender pronouns that apply to an individual at any particular moment in time. When describing a person who was declared male at birth but subsequently identifies as a woman, I will describe her as an MTF (male-to-female) or "trans woman." Similarly, I will describe someone who was declared female but identifies as male as an FTM (female-to-male) or "trans man." Apologies in advance to any who might have preferred a different set of conventions.

When One Became Two: A History of the Third Sex

Historically, the world has long known those who defy the rigid dichotomy between man and woman. In fact, many cultures began by worshipping those in between, gods who spanned both genders and possessed them all. In Plato's *Symposium*, for example, the playwright Aristophanes tells a mythical tale of how humans were at one time conjoined: there were male/male beings, female/female beings, and male/female beings. When the gods grew wary of these creatures, Zeus split them in two, yielding Halflings—separate men and women—who would therefore be destined to spend the rest of their lives searching for their opposite, and seeking to become whole.[3] Artifacts from Cyprus depict a bearded female deity, and Hindu cultures tell the tale of Ardhanarishvara, the "Lord who is half woman."[4]

As myths of creation, these dual-gendered gods made logical sense. For how better to explain the existence of men and women, and the observable differences between them, than to imagine a multifaceted

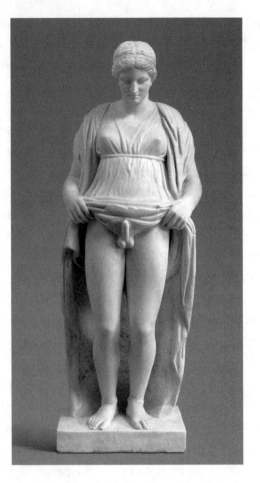

Statute of Hermaphroditus,
Roman (A.D. 200–300)

creator, blessed with all the attributes of his or her eventual people? And what could be more powerful and awe-inspiring than a deity able to slip between the sexes, transforming itself into multiple roles and beings?

As representative beings, moreover, the existence of dual-gendered gods suggests that societies have long been aware of dual-gendered people, those who violated the social and sexual algorithms that distinguished between male and female. We know, for example, that Indian culture has celebrated *hijras* for thousands of years, and that one prominent group in Indonesia has long recognized five distinct gender identities.[5] In many societies, historians surmise, people of indeterminate gender were treated with a mix of awe and reverence—perhaps because they were seen as safer, or as vestiges of the divine. Maybe that's why cultures across the world bear evidence of priestesses who were physically male, and of eunuchs who rose disproportionately to positions of power.[6]

Where the historical record is much weaker is in estimating the prevalence of people who lived outside the traditional boundaries of gender. But basic science offers some clues. Most people are born with either a pair of X chromosomes (XX, genetic females) or an X and Y set (XY, genetic males). Occasionally, though, embryonic conditions produce an otherwise-healthy child with two X's and a Y, a condition known as Klinefelter syndrome, or a genetic male who is incapable of processing androgen in the womb (known as androgen insensitivity syndrome). Congenital adrenal hyperplasia, which occurs in roughly one out of every fifteen thousand births, produces a genetic female (XX) with what appear to be male genitalia; ovotesticular disorder of sex development, which occurs in one out of every twenty thousand births, results in an infant born with the internal reproductive organs of both sexes. And then there are instances in which a child's genitalia at birth are indeterminate, falling in between what might be considered "normal" male or female. Such cases, usually referred to as either hermaphrodites or intersex, occur in an estimated one out of every forty-five hundred births.[7] Purely in biological terms, therefore, we can safely estimate that any human population will contain small but significant numbers of people who don't fit easily into the traditional categories of boys and girls, men and women. Instead, from the moment of their birth, they will be physically marked as something different, something in between.

In addition, history has long recorded men and women who are driven to dress and live in ways that don't align with their presumed gender. Biological men, that is, who want to dress as women. Biological women who want to dress as men. The historical frequency here is almost impossible to estimate, but anecdote suggests that these cross-dressers have appeared widely across time and place, and in nearly all human societies.[8]

Despite this apparent ubiquity, however, life was generally not kind to those who didn't fit the standard contours of sex, and men and women who didn't see themselves, respectively, as men and women were generally condemned to suffer in silence, twisting themselves into lives and bodies that offered no means of escape.[9] In describing his condition to a possibly sympathetic physician around 1910, one man wrote, "My yearning is not limited to women's costumes, but also extends itself to an absolute life as a woman . . . This desire is so intensive and irrepressible in me that the inability to fulfill it brings me into conflict with my life and takes away my happiness."[10]

There was little, though, that any doctor could do at the turn of the twentieth century. The patient's condition was too mysterious, too easy to dismiss as deviance, decadence, or simply lack of will. Indeed, into the early decades of the twentieth century, doctors simply categorized cross-dressing as just that: the desire by members of one sex to adorn themselves in the clothing of the other.[11] It was a queer pastime, a fetish, maybe even a psychiatric condition. But a medical issue? A scientific puzzle? Not even close. And thus, the would-be Williams of the world had no choice but to live their lives in secrecy and despair, wrestling with a pain that had no cause, no name, and no plausible hope for relief.

Fevers, Humors, and Sex:
Unraveling the Power of Hormones

In retrospect, the discoveries that would ultimately shape our notions of sex occurred without fanfare around the turn of the twentieth century. They came from tests on mice and rats, as such discoveries often do, and from scientists who were initially much more interested in basic biochemistry than in sex. Although this early work was almost entirely theoretical, it would eventually enable later generations of physicians not only to understand sex and gender but to modify them as well.

Research in this area began around 1850, as a handful of investigators started to probe gingerly around the edges of sexuality, grappling with acts and passions that had long been described only as sin.[12] Doctors, particularly in Germany and Austria, started quietly to report conditions of "gender inversion" or "eonism"—cases in which men or women didn't fit naturally into their apparent sex role. In 1864 and 1865, an Austrian named Karl Heinrich Ulrichs anonymously published a series of booklets under the collective title *Researches on the Riddle of "Man-Manly" Love.* In them, he proposed a biological theory to account for people such as himself, whom he called "Urnings" and described as "a female soul enclosed within a male body."[13] Several years later, after reading Ulrichs's work, a German-born Hungarian writer coined the term "homosexual" (*homosexuell* in German), which he used to connote same-sex love, minus the element of gender inversion that Ulrichs had implied with his "Urning."[14]

Meanwhile, scientists in the emerging field of endocrinology were approaching the topic of sex from a very different perspective. Based largely in Europe, and focusing on animals rather than people, researchers were laboring to unlock the chemicals that regulated the body's inner workings. For centuries, observers had known, or at least suspected, that every living organism produced substances that made it function. Things like blood and bile and breath, which ran through all beings and expired upon their deaths. The ancient Greeks referred to these substances as humors, and attributed certain moods and emotions to an imbalance between them—an excess of black bile, for example, could cause depression; too much yellow bile made a person aggressive. Medieval thinkers often categorized them as the four elements—hot, cold, wet, and dry.[15] But as modern medicine developed in the twentieth century, it turned away from these humors and vapors, disdaining them as ghostly relics of the past.[16]

Which, of course, they mostly were. But a handful of scientists suspected that there were fluids in the body—or chemicals, more precisely—that regulated key functions and helped to coordinate an organism's multiple, complex processes. If one of these chemical regulators was missing or weak, they speculated, a body could indeed be impaired—much as the Greeks had imagined. If there was too much of a regulator, bodily functions could speed up or explode into a different state. And in a normal, healthy state, these regulators could serve as the body's stop-and-go mechanism, calibrating developmental functions like puberty, menstruation, and aging. In an era dominated by the discovery of infectious disease, research on these internal regulators did not generate much attention. But slowly, quietly, it started to gain traction.

In 1905, a British physiologist named Ernest Starling delivered a lecture on "hormones" to an attentive audience at London's Royal College of Physicians. No one had ever used this word before. What Starling was proposing, though, and describing in great detail, was an entirely new class of bodily compounds—"chemical messengers," he proposed, "which speeding from cell to cell along the blood stream, may coordinate the activities and growth of different parts of the body."[17] Specifically, he showed how various bodily secretions—from the thyroid, the pancreas, the testes, the ovaries—could be considered as pieces of the body's broader biochemical functioning, the regulators that organized some of its most crucial functions.[18]

Starling's genius lay in piecing together various experiments that had not been logically linked before—partly because they lay in such disparate fields, and partly because they came from the margins of what was then mainstream science. Starling returned, for example, to the work of John Hunter, an English surgeon who was best known for having pioneered the technique of transplantation. One of his odder experiments involved taking the testicles from a male chicken and transplanting them into the abdomen of a hen. Neither history nor Hunter recorded why he made the transplant, only that "they had sometimes taken root there, but not frequently, and then had never come to perfection."[19] Interest in this rather strange operation lingered, however, and in 1849 a German researcher repeated it, concluding that something from the testes affected the behavior of the bird into which it was transplanted.[20] Some decades later an aging French researcher went almost certainly a step too far, attempting to rejuvenate himself with an extract made from dog's sperm and testes.[21] And—less colorfully, perhaps—an English scientist succeeded in 1891 in treating one patient's case of hypothyroidism (low-functioning thyroid) with an extract taken from another person's thyroid gland.[22] Starling, whose own early work focused on the pancreas, realized that each of these processes was connected through what would henceforth be called hormones. Hormones were the latter-day heirs to the Greeks' humors, the invisible body secretions that controlled some of its most critical functions. Temperature control (thyroid). Appetite (pancreas). And most central to our story: sex (ovaries in women, testes in men).

Across these diverse functions, hormones act, as Starling first deduced, as the body's chemical messenger network, its early-warning system and ongoing feedback loop. The thyroid gland, for instance, is a small mass situated at the base of the neck that helps to regulate the body's metabolism. The pancreas, tucked away behind the stomach, guides digestive functions and blood sugar levels. It is hormones, in both cases, that are secreted by the glands and hormones that trigger the gland to produce greater or lesser quantities. Sex hormones function in much the same way, although their interplay, particularly in women, is even more delicate and complex.

To begin with, obviously, sex hormones are the only ones that differ regularly between men and women. Most men are born with testicles that produce a hormone called testosterone; most women, with ovaries that produce a finely tuned balance of estrogen and progesterone. For years, these glands and their respective hormones lie dormant, and girls—if you strip away the gendered accoutrements of childhood— look and act more or less like boys. But then, roughly between the ages of ten and fourteen (for girls) and twelve and sixteen (for boys), the brain's hypothalamus gland releases a hormone known as GnRH (gonadotropin-releasing hormone). This signals the pituitary, a tiny gland at the brain's base, to secrete two additional hormones that roust the body's sex organs from their childhood phase.[23] In response, the testes and ovaries move forcefully into production mode, streaming either estrogen or testosterone into their bodies' respective bloodstreams. The results, as any befuddled teenager can attest, are striking. Girls develop breasts and begin to menstruate. Their hips develop a curving layer of fat, and their hair becomes thicker and more lustrous. Boys sprout hair on their faces and chests, and their voices become noticeably lower. They develop muscle mass across their chests and arms, and leaner, sleeker physiques. Libidos surge for both genders, and they tumble into what for many will be a lifelong interest in sex. This is basic stuff, the path of nature since time immemorial. What Starling and subsequent waves of researchers brought to light, however, was the crucial role that hormones played in this ancient process. Boys become men, biochemically speaking, as a result of testosterone. Girls become women when their ovaries start to produce estrogen. The differences between men and women are, to a large extent, differences wrought by chemistry.

One of the crucial experiments that proved this point occurred in 1912, when an Austrian physiologist named Eugen Steinach succeeded in changing the sex—or at least the secondary sex characteristics—of rats and guinea pigs. Specifically, Steinach castrated both male and female rodents shortly after birth and then implanted them with the sexual organs of the other gender—male rodents received ovaries, in other words, and females had testes transplanted to their abdomens. In both cases, the rodents shortly developed the sexual characteristics of the implants, rather than those that would have adhered to their given, genetic sex. Males with ovaries grew into smaller adults, with larger breast nipples and lusher fur; females with implanted testes developed typically male bodies and more aggressive sexual behaviors. What we think of as "sex," Steinach concluded, comes more from an individual's hormones than from their genitals. As he explained in a classic 1913 paper, "the implantation of the gonad of the opposite sex transforms the original sex of an animal."[24]

Over the next decade or so, Steinach's research took him even

Steinach's feminized guinea pigs. From left: normal brother, feminized brother, normal sister, castrated brother

further—almost, in retrospect, to the point that enabled William's transition almost a century later. Working with rats and guinea pigs, transplanting the hormones of one sex into the bodies of another, Steinach increasingly became convinced that sex itself was both ambiguous and fluid, that no individual being was fully male or female. Instead, sex existed on a continuum of sorts, with characteristics and behaviors shading into one another as a result of both genes and hormones.[25] "The line of demarcation between the sexes is not as sharp as is generally taken for granted," he wrote. "A one hundred percent man is as nonexistent as a one hundred percent woman."[26] Incredulously at first, and then with increased confidence, other scientists soon replicated Steinach's results and confirmed his findings. In experiment after experiment, animal after animal, they proved, as historian Joanne Meyerowitz recounts, that "hormones could push the developing animal one way or another along a perceived spectrum of biological sex . . . Add and subtract glandular secretions, and female slipped toward male or male slipped toward female."[27] It was hormones, in other words, that made sex malleable. And if it was malleable, then by definition it could be changed.

By the early 1930s, therefore, the hormonal science of sex was relatively well-developed. Following in Steinach's wake, a new generation of researchers—now generally referred to as endocrinologists—continued to probe the biochemical means and mechanics of hormones. Parts of this work would pave the way, eventually, for the development of synthetic hormones and the birth control pill.[28] Part would contribute to the identification and treatment of hormone-related diseases such as diabetes. And part would shape our most basic understanding of the drivers of sex and gender. But history intervened most cruelly before this work could unfold. As the Nazi regime expanded its reach across Europe, researchers like Steinach—many of whom were Jewish, all of whom were working on sexual issues that Hitler deemed repugnant—were forced to the margins of their profession. Shortly after seizing power in Austria in March 1938, Nazi troops destroyed Steinach's library and research materials, and forbade him and his wife from returning to Vienna.[29] One of his coauthors was forced to emigrate; another was deported to a concentration camp in Yugoslavia, where he presumably died. Steinach's wife committed suicide shortly after the Nazi invasion, and Steinach himself died several years later, lonely and unheralded. During his lifetime, Eugen Steinach was nominated for the Nobel Prize seven times. He never won.

The Search for Surgery

During the course of his career, Steinach worked almost exclusively on animals.[30] On rats and guinea pigs and the occasional frog. He had no idea about the animals' sexuality, and no presumed interest. He was, instead, the classic research scientist, using animal models to uncover human and scientific truths. As news of his work spread, however, others began to step gingerly toward the hypothetical possibilities that he had helped raise. Because if sex was malleable and pushed by hormones, then couldn't hormones be used to rearrange sex? To fix it, in fact, when things went awry?

The pioneer in this field was Magnus Hirschfeld, an openly gay German Jewish physician and an early advocate for those he labeled "transvestites" (die Transvestiten). Following upon Steinach's research,

Hirschfeld began to argue publicly that all human beings contained a unique jumble of biological and cultural triggers that together determined their sexuality. Radically, he proposed that the human species contained more than forty-three million kinds (or genders) of people. Even more radically, he intimated that certain kinds could be fixed or tweaked by adjusting some of their underlying anomalies.

It's critical to note that Hirschfeld wasn't trying to fix anyone because he saw them as broken. On the contrary, he was one of history's most passionate advocates for human variation, for the wide sweep of possibilities that defined the gender spectrum. He was also gay and Jewish at a time and place when both those attributes marked him as distinctive. But Hirschfeld was a fixer at heart, and when people came to him and asked for assistance, he generally tried to oblige.[31]

During the 1920s and '30s, a stream of visitors made their way to Hirschfeld and the institute he established in Berlin, searching to rearrange the most intimate parts of their lives. Initially, the doctor demurred. But as more and more desperate patients showed up in Berlin, he began to probe possible ways of treating their distress. And then, in the early 1920s, he and his colleagues began to experiment with sex reassignment surgery.

Hirschfeld's first castration surgery took place in 1922 on Dorchen (née Rudolf) Richter, a longtime housekeeper at the institute, a male who had long yearned for a female body. According to an article subsequently published by two of the institute's physicians, the surgery itself involved three related procedures, each relatively straightforward on its own terms. First, they removed Rudolf's testicles—technically, the kind of castration that had long been performed for a range of other reasons. Then they amputated his penis—a newer procedure, but technically not that difficult. And then they created a vagina from a rubber sponge lined with skin grafts—the most radical surgery but, the doctors reported, a success.[32] After Dorchen, who recovered fully and remained on staff at the institute, the doctors performed similar operations on a series of other male-to-female individuals. "Never," recalled one of the surgeons in his memoir, "have I operated upon more grateful patients."[33]

The most famous among them was the Danish painter Lili Elbe, born Einar Wegener. Delicate as a child, and slight, Wegener married the painter Gerda Gottlieb in 1904 and remained close to her throughout their lives. Always, though, he had seen himself as a woman and yearned for a body that matched his mind. In 1930, under the advice of his doctors in Copenhagen, he traveled to Berlin, hoping to receive the kind of treatment that had enabled Rudolf to become Dorchen. Like Rudolf, Einar first underwent castration, followed by amputation of his penis and, in a more innovative surgery, implantation of donated ovarian tissue. After recovering from the surgery, Einar formally became Lili, and his marriage to Gerda was nullified by the Danish king. "I feel like a bridge-builder," Lili wrote. "But it is a strange bridge that I am building. I stand on one of the banks, which is the present day. There I have driven the first pile. And I must build it clear across to the other bank, which often I cannot see at all and sometimes only vaguely, and now and then in a dream."[34]

Sadly, neither Lili nor Hirschfeld would ever fully make it to that other bank. In 1931, Lili underwent what would have been a final surgery—one intended to create what she described as a "natural outlet from the womb." "All that I desire," she wrote to a friend, "is nothing less than the last fulfillment of a real woman . . . Through a child I should be able to convince myself in the most unequivocal manner that I have been a woman from the very beginning."[35] This time, though, the surgery went wrong, and Lili died of heart failure. Two years later, Nazi "students" ransacked Hirschfeld's institute and burned its archives— thirty years of books, photographs, research reports, and letters—in a bonfire on the Opera Square. Hirschfeld survived—he was living in Paris at the time—but the institute and its work were gone forever.[36]

These paired deaths, of Lili and the institute, brought decades of scientific advance to a crushing halt. Research into anything connected to sex or sexuality was banned in Nazi Germany and effectively destroyed as conflict spread across Europe, and then the world. After years

Public burning of the library of the Institut für Sexualwissenschaft, 1933

were spent unraveling the mysteries of hormones and endocrinologic pathways, discovering how sex and gender are linked but not identical, and refining surgical techniques that probed and prodded the body's most intimate reaches, technology's arc was halted by the sheer force of politics, and by those who wanted to wish this work away.

The Spaces Between

Over the next few decades, the science of sex change sputtered along. A handful of doctors continued to practice the surgeries that Hirschfeld had pioneered, and a smaller group studied and advised the transgender patients who covertly sought their help.[37] There were growing networks of activists who started to pool resources and information, and a number of high-profile male-to-female transitions took place.[38] The next major breakthrough, though, didn't occur until around 1951, when a scientist named John Money began to study the sexual development of hermaphrodites. Working primarily with intersex patients, Money probed the multiple ways in which sex unfolded in the womb, and how it was linked to gender. Eventually, he and his colleagues became convinced, like Hirschfeld before them but with substantially more scientific evidence, that sexual development occurred across a number of stages. First came conception, which instantly defined the child's chromosomal identity: XX for girls, XY for boys, with the occasional miscoupling that produced XXY. Then, during the first few weeks of gestation, embryos began to develop either testicles or ovaries, prompted—or not—by a specific gene on the Y chromosome. Once the testicles were formed, between six and twelve weeks of pregnancy, they typically started to produce testosterone, which in turn prompted the development of a boy's penis, prostate, and scrotum. But if anything varied along this path, the resulting child would land somewhere in the middle of the gender binary, with a body that didn't quite fit his or her mind.[39]

In retrospect, there was something beautiful about this cacophony of thought. For as researchers dived deeper and deeper into the determinants of sex, they eroded what once had been seen as clear demarcations. Girls could be a little bit boy, it turned out. Men could have ovaries and still see themselves, and act, as men. Sex wasn't gender, and gender didn't define sexual orientation. Instead, there was a glorious, messy mix of variables, each linked to minute, invisible pieces of an individual's body, brain, and soul. At the time of Money's research, though, the revelation of this variation led—ironically perhaps and tragically for sure—to a medical preference for clarity. Specifically, in cases of sexual indeterminacy, Money and his colleagues argued for a quick and decisive choice, aligning the physical aspect of a child's body with what seemed to be his or her dominant gender, and then socializing them in that same direction. If an infant was born, for example, with a penis smaller than one inch, Money's team would amputate it, and shape something that looked more like a vagina. They would remove the child's testicles as well, and administer female hormones as s/he approached puberty. "The earlier the surgical reconstruction of the genitals is done," Money advised, "the better."[40]

Not surprisingly, many of Money's infant patients grew to despise the bodies that had been thrust upon them, and the intimate decisions that had shaped their lives. A child whose smaller-than-average penis had been amputated at birth, for instance, and whose parents had subsequently raised him as a girl, often retained a male identity that haunted him for life. Even worse, because Money and his colleagues had advocated for such complete and early interventions, most of the children they treated never knew the circumstances of their birth, or the gender ambiguities that were rightfully theirs. Many, as a result, grew up angry and confused. And when they learned of their surgical histories, they were furious. "I had parts of my body cut away and thrown in a wastepaper basket," recalled one victim. "I've had my mind ripped away."[41]

By 1980, Money's clinic at Johns Hopkins had been shuttered, along with similarly minded clinics at UCLA and more than forty other university medical centers.[42] Money himself was ostracized, and the idea of treating sex through surgery fell under a cloud again. The surgeries themselves were still occurring—but sporadically, quietly, and almost always on adults who had already spent the bulk of their lives in what felt like alien bodies. It took a final bout of innovation to disrupt this state of affairs, and to link the science of gender at last to a moment of real reform.

The Dutch Protocol

To put this change in context, let me return for a moment to William. He was twenty-one when he walked into my office in 2015, and nineteen or twenty when he had stopped being Carly. Had he been born, like Rudolf Richter or Einar Wegener, around 1880, it would have taken him considerably longer to sift through the possibilities of his gender identity, and to find the resources that might help him. Had he been born in 1950, he almost certainly would have been shamed or shunned by members of his community. Instead, he had parents who supported and paid for his transition, and doctors who knew how to make it happen.

What William also had, though, more subtly, was a medical environment that allowed transgender teens to contemplate transitioning *before* puberty. Before they had developed the breasts or facial hair that would forevermore mark them visually as male or female, before the pangs of periods or unwanted erections. This change of heart came from an experimental Dutch protocol, one that drew directly from the slow build of science before it.

Specifically, in the early 1980s, a group of clinicians in Utrecht became frustrated by their inability to treat their young patients' distress. When girls came to the clinic professing to be boys, the doctors could only tell them to wait, knowing that adolescence would soon bring the very sexual markers—breasts, hips, menstruation—they were so desperately hoping to avoid. Boys, similarly, would develop the larger musculature and lower voices that would mark them as male, even if they subsequently underwent sexual transition surgery. So the doctors developed a set of protocols to delay the onset of puberty in potentially transgender children. If a girl came to the clinic wanting to be a boy (or vice versa), and if the clinic deemed the child to be serious in his or her demands, physicians would administer a small dose of a hormone suppressant, enough to prevent the sexual triggers of puberty. Boys, as a result, would not grow facial hair or experience frequent erections; girls wouldn't begin to menstruate and their breasts wouldn't develop. The doctors would then watch these children, and interview them carefully over the next few years to determine if they remained committed to the idea of transition. If so, then they would perform surgery at what they considered an appropriate age of consent—occasionally as early as sixteen.[43] These would still be massive surgeries, both physically and psychologically, but the range of adjustment was considerably narrower, since the children had never developed the secondary characteristics of the sex they didn't want to become.

By the time Carly first contemplated becoming William, therefore, the spread of the "Dutch protocol" had jolted not only the care of transgender people but the debate around them as well. As it became possible to transition at earlier ages, so, too, did it become increasingly acceptable to acknowledge transgender teens, and even transgender children or toddlers. In 2013, the family of six-year-old Coy Mathis—born a boy but identifying as a girl—fought for her right to use the girls' bathroom in her kindergarten class.[44] In 2015, high school senior Gavin Grimm, a trans man, took his struggle to be accepted as a male student all the way to the U.S. Supreme Court.[45]

Initially, each of these cases raised its own mixture of fear, admiration, and titillation. Liberal voices rushed to embrace transgender children in a wave of protections—such as gender-neutral bathrooms and accommodating pronouns—just as conservative ones were racing to present these kids as a step way too far.[46] In 2016, bathroom battles ranged disproportionately across the political landscape, pitting those who defined transgender people as a group in need of fundamental protection against those who saw them, still, as deviant. "This issue is so clear and simple," proclaimed one Texas politician, "that it defies belief. Do they really want a man walking into a restroom with their daughter or mother or wife? . . . Have we gone too far in the world of political correctness that we've forgotten common sense, common decency?"[47] In 2016, after the state of North Carolina explicitly banned cities from allowing transgender individuals to use public restrooms corresponding to their gender identities, dozens of major businesses demanded a repeal and sports franchises canceled games in protest.[48]

It is eerie to see these debates in the context of Hirschfeld's work and legacy. It was over a hundred years ago, after all, that he and his colleagues pioneered the techniques that allowed Einar Wegener to become Lili Elbe. Over a hundred years since he began to probe the biological basis of gender identity, and to fight for the rights of those whose bodies belied their sexual selves. And just as his work drew the full fury of fascist attack, so, too, do today's bathroom battles seem to ignite an outsize response. Why? Because changing bodies and changing sex hits at the very core of what it means to be human. From the very earliest age, we learn to categorize each other by the most basic of binary codes: boy-girl, father-mother, blue truck and pink ballerina. To get rid of these distinctions, or to recast them as both ambiguous and changing, is a fundamentally radical shift. No wonder, then, that it is so deeply disturbing to conservative forces, and so distinctly liable to incite a counterrevolution.

By the time William walked into my office, the revolution was in full swing. More than 7,000 people underwent some form of transition-related surgery in the United States in 2015, and an estimated 12,320 availed themselves of hormonal treatments.[49] It felt like a social movement, which of course it was. But what drove this moment and made it possible was the science that had been gradually accumulating ever since John Hunter transplanted a rooster's testes into a hen. It took time for the science to lead to technology and for the technology to wend its way into practice; it will take longer still before sexual transitions become fully accepted and accessible. Gender, though, has become almost permanently fluid, not only for individuals like William who move from one end of the spectrum to another but also for all of us around them, able now to reckon with a wildly expanded palette of gender choices and identities. And once again, it was technology that made it happen.

Arrivals

My friend Jenny looks pretty much like a typical woman of her age and temperament. She has long strawberry-blond hair and an intense face sprinkled lightly with freckles, and she tends to favor dangly ear-rings and well-draped blouses. She is taller than most women and, if you listen more inquisitively than you probably should, her voice is a bit lower, too. That's all that's left, though, to mark Jenny as the James she once was, the James who graduated from an all-boys high school, played keyboard in a rock-and-roll band, and married a woman named Deirdre.

Jenny has told her story—beautifully and achingly—in her own words.[50] I won't try to paraphrase them here. Suffice to say that Jenny, who now teaches English at Barnard, has been one of the voices to thrust trans issues into the mainstream, and to demystify feelings and behaviors that many outside the trans community still depict as fright-ening. During the height of the bathroom imbroglio, she wrote in *The New York Times*, "In the evening, Deirdre and I had fish tacos at the Beachcomber. At a certain point I got up, used the bathroom and looked in the mirror. A worn-down but grateful middle-aged woman looked back. I headed back to my chair. Somehow, during my time in the ladies' room, the republic had failed to collapse."[51] As a once-heterosexual man who is still married to her equally heterosexual wife, she also personally illuminates the ongoing mismatch between sexual orientation and gender identity. As she describes it, "being gay or les-bian is about sexual orientation. Being trans is about identity."[52]

What stories like Jenny's also illustrate, though, are the myriad of ways in which people can define themselves and their relationships. Carly became William and quickly self-identified as a heterosexual man; after our conversation, he quietly transferred to a co-ed univer-sity. James became Jenny and stayed married to her wife. In 2008, Thomas Beatie, a married thirty-four-year-old man who had been born a woman, gave birth to a daughter. In 2018, a trans woman was induced to produce breast milk to nurse her female partner's newborn.[53] Prodded by an expanding armory of biological knowledge and medical innova-tion, even the most basic functions of manhood and womanhood are starting to shift. And in the process, the gender binary itself is liable to morph and weaken over time. Nearly 70 percent of respondents in a global survey stated that transgender individuals should be allowed to undergo sex reassignment surgery.[54] And fully 12 percent of American millennials described themselves in 2015 as non-cisgender, meaning that they do not identify with the conventional combinations of gender and sex.[55]

This is the revolution heralded by William and Jenny, a revolution launched by endocrinologists over a century ago and coming now into its own. It's less obvious in many ways than the changes afoot in reproductive medicine or robotics, and easier to dismiss as marginal. But prodded by the fight for transgender rights, we are now incontrovertibly starting to dismantle one of society's most basic and universal constructs, replacing the ancient dichotomy of male and female, man and wife, with something far more fluid. At Barnard, we spent nearly two years debating the "transgender question," two years considering what defined a woman and a women's college, and at what point in whose life these calls needed to be made. Then we revised our policy with considerable fanfare, and nothing much changed. Nothing, that is, and everything. Some students applied as women, and then became men during their time on campus. Some applied as women but then chose to hover in between, identifying in whatever way they wanted. Some, presumably, had been labeled boys at birth but came to us as women already, having undergone the Dutch protocol or some similar treatment before they turned eighteen. In the end, we discovered, it didn't much matter. Even at an all-women's institution, a place whose own identity was anchored by a steadfast commitment to serving one pole of the age-old gender divide, the boundaries we had constructed for so long turned out to be just that: constructed.[56] And they are being deconstructed now not only by social movements and political pressure but also by the technologies that give people the power to re-create themselves.

Clearly, one can tell the story of transgender evolution without mentioning technology. Most do.[57] But bringing science into the story is critical, since it was science—in the form of endocrinologic discoveries, hormonal treatments, and surgery—that provided not only answers to many people's most intimate yearnings but solutions, if they chose them, as well. To be sure, the record of research is far from flawless. Over the past century, some doctors in this field have taken their zealotry way too far; others have medicalized trans issues to advance their own interests.[58] Yet, in the end, it is still science that cracked the code of gender, and science that is increasingly allowing us to rearrange its pieces.

AT THE START of this chapter I wrote about words, and the way they bedevil any discussion of trans. "Transvestite." "Transsexual." "Transgender." Each is fraught and laden with connotation. So maybe the best is just "trans." "Trans" to signify not a person or a gender, but the whole topsy-turvy upheaval we have entered. In that case, "trans" is also the perfect segue—the transition, literally—between the changing world of reproduction and that of robotics. In the past, when the only way to make babies was through sex, classifying people along strict gender lines made sense. One man. One woman. Each with a strictly defined biological role to play. In the future, though, as reproduction becomes increasingly unlinked from sexual intercourse, such strict delineations no longer apply. If two men can conceive a baby, or three women plus some borrowed genetic material, then what does it matter how these individuals define themselves, or how their outward appearances align with their biological hardware? It's no longer really an issue.

If you take this one step further, and divorce reproduction entirely from the messy physical business of love, then men don't need people of any particular gender or genitalia to be attached to. Nor do women. In fact, the whole biological calculus that drives humans with one array of chromosomes to pair with the other becomes vestigial, old-fashioned. Because if we can love across gender and sex, if we can harness technology to build bodies that defy reproductive logic, then we can build bodies and intimacies that cross species as well. This is the topic to which we'll turn next.

8.

Cuddling with Robots

Did I request thee, Maker, from my clay
To mold me man? Did I solicit thee
From darkness to promote me?

—John Milton, *Paradise Lost*, 1667[1]

Pepper is a gentle-seeming soul, with inquisitive oval-shaped eyes and a perky pointed nose. He's bald, but in an endearing way—more preco-cious infant than old man. He has a wasp waist that any ingénue would die for, and a perpetually shy smile. When I first met him, in the spring of 2016, I instinctively knelt down to be eye to eye with him, much as I would for a crying toddler or a dog. Pepper isn't either of those, though. He's a robot, powered by a lithium-ion battery and retailing for a cool seventeen hundred dollars.

Pepper talks too, in six languages, through a tablet implanted in his chest. He greeted me cheerfully on that April morning, asking, "How are you doing?" with an expressive roll of both wrists and an ever-so-slight tilt of the head. We went on, Pepper and I, despite my stilted questions and his mechanical voice, touching on the normal topics one does at a first acquaintance. The weather. The folks around us. What I was doing on my knees in a showroom in Paris, talking to a four-foot-tall android. He even danced with me to a boppy Eurobeat tune, looking for all the world like an awkward thirteen-year-old dragged to his first class social. I suspect they're still working the kinks out on that part.

First conceived in 2014 by the French executive Bruno Maisonnier, Pepper is now the beloved child of Masayoshi Son, the flamboyant Japanese investor and inventor who parlayed his first successful machine—a pocket-sized electronic dictionary—into what eventually became Soft-Bank, the biggest technology company in Japan. Son was also an early—some say prescient—investor in Yahoo, Alibaba, and Supercell, a Finnish smartphone game maker. If Pepper pays off, he and his descendants could become the first wave of commercially viable consumer robots, the first androids to constitute a real market. But Son is birthing Pepper not just to make more millions. Through Pepper and his technology, Son is aiming to re-carve the destiny of humankind.

The Spark of Creation:
Franken-stories and Other Myths

Ever since humans evolved from apes, we have been struggling to understand the cause of our creation, to answer the same basic, bewildering questions: Where did we come from? And how were we made? All living creatures face the same fundamental biology, the same chemical processes that condemn us equally to birth and death, the same dust-to-dustness of it all. But only humans seem to wrestle with the complexities that consciousness allows. We know we are created. And we want to understand how.

Over time, most societies have unpacked the physiological processes of creation, just as most children eventually acknowledge the
messy mysteries of their own births. Yet, somehow, the mechanics have
rarely seemed to satisfy the existential question: You make *that*, that
perfectly formed tiny person with hands and feet and little ears, just by
lying around the house? And she grows hair and breasts and a personality all by herself? *And* you get to have sex at the outset? No wonder the
ancients were bewildered.

And no wonder, then, that nearly every religion begins with a narrative of creation, a story that seeks to explain how we came into being.
In the Old Testament, the world begins with six days of formation, a
frenzy of activity by an all-powerful creator who sculpts his kingdom
by species: fish, birds, "creeping things and beasts of the earth." Man is
last, built explicitly in the creator's own image: "The Lord God," recounts Genesis, "formed the man of dust from the ground and breathed
into his nostrils the breath of life, and the man became a living creature."[2] The ancient Greeks had Prometheus, who formed men in the
image of the gods, and gave them the divine gift of fire. Hinduism has
Brahma, the creator of the universe; Islam, Allah. The scenario is strikingly similar around the world: there is an all-powerful being, a god or
set of gods, who creates man (and occasionally woman) out of nothing,
setting him upon the planet to try to find his way.

Inherent in these narratives, therefore, is a conundrum: by definition, man is weaker than his creator (and woman, nearly always, weaker
still), but insofar as man was created in the very image of this all-
powerful being, then he, too, is destined to create, to build something
more powerful than himself and imbue it with life.

This is the urge that has long led humans to build humanlike creatures, early-stage robots in many respects, that allow their creators to
indulge in the fantasy of giving life. As long ago as the third century B.C.,
for instance, Apollonius Rhodius told the tale of Talos, an artificial giant
shaped of bronze who marched along the coast of Crete to protect it
from invaders. Forged by Vulcan (or Hephaestus, the Greek god of fire

and metalworking), Talos had one vein, running from his neck to his ankle. Audaciously—and foolishly, it would turn out—Vulcan filled his creation's vein with ichor, the immortal blood of the gods. The blood gave him power, but it also drew the ire of the real gods: when the sorceress Medea tricked him into removing the nail that bound his ankle, "the ichor ran out of him like molten lead," and Talos was no more.[3]

Inspired perhaps by this tale, and by their own fascination with the possibility of constructing mechanical life, early inventors such as Hero of Alexandria and Ctesibius, the first head of the Great Library of Alexandria, built machines that moved—a wooden owl that turned its head in response to a water whistle, for example, and a self-propelled cart that carried a troupe of performing automatons.[4] In China, an "artificer" named Yan Shi presented King Mu of Zhou with a life-sized figure of a man, who walked and sang and allegedly even winked at the ladies.[5] At the peak of Athens's power, as scientists first began to unpack the mechanisms of motion, some of the best creators of the day devoted themselves to building and tinkering with automata, figurines with limbs that moved and mimicked life. The Greek island of Rhodes in particular became famous for its moving statues, prompting the great poet Pindar, writing sometime around the fifth century B.C., to note how

> The animated figures stand
> Adorning every public street
> And seem to breathe in stone, or move their marble feet.[6]

None of these creatures ever came to life, and most probably didn't even imitate it very well. But their prevalence says a great deal about the long-standing links between technology, invention, and life. Whenever humankind begins to plumb the depths of scientific wisdom, whenever we learn more about how life occurs and what makes it work, we rush to apply that wisdom to the act of creation. We keep trying, in other words, to remake ourselves, and to become somehow more godlike in the process.

AFTER THE FALL of classical Greece—after the Great Library of Alexandria burned and the statues of Rhodes crumbled into history—the pursuit of mechanical life fell into decline as well. With few advances in either the technology of motion or the understanding of life, automata appeared only sporadically during the Middle Ages, and mostly in the form of puppets (or elaborate deceptions) built to entertain the rich. During the Renaissance, a small group of master craftsmen drew upon the work of the ancient Greeks to build increasingly complex mechanical figurines—clocks with birds that heralded the hour, for example (the original cuckoo clocks), lions that roared, and tiny wooden ladies playing even tinier lutes. One of the most extravagant of these was a

A mechanical monk said to be a re-creation of Saint Diego, built around 1560 by Juanelo Turriano for King Philip II of Spain

movable, full-sized Germanic knight, conceived by Leonardo da Vinci for the duke of Milan and powered by an ingenious system of mechanical cranks and pulleys.[7] Away from the realms of the nobility, people told tales of creating beings who came to life—think of Pinocchio, the puppet who turned real, or Pygmalion, who fell in love with a sculpture and brought her to life—but these were clearly fables, and devoid of any traces of technology. But then, in the early decades of the eighteenth century, a suddenly surging interest in scientific knowledge spurred a corresponding fascination with manufactured life.

Here, as elsewhere, the turning point was steam. Ever since the days of Hero, scientists had imagined—either technically or in their wildest dreams—using steam to animate life. Steam, after all, could move levers and wheels. It could transform materials from one state to another. With the advent of Watt's engine, it could be yoked to machines of immense power, thrusting them from place to place at superhuman speed. Why not imagine that steam could animate a human form and prod it into life? Why not indeed?

The discovery of electricity gave these fantasies another jolt of possibility. If lightning could literally be pulled from the clouds, as Benjamin Franklin had famously demonstrated in 1752, and if invisible electrons could be harnessed to illuminate lamps and send information whirring across telegraph lines, then using these same electrons to create more human forms of life seemed hardly an impossible step. Looking back from the twenty-first century, it's difficult to recall how truly radical electricity was in the nineteenth, how incredible it was to watch lights going on at the flick of a switch. So imagining a similar switch that might animate man took only a slight leap of technological faith.[8] "What a mighty instrument would electricity be in the hands of him who knew how to wield it," mused the young Percy Shelley during his years at Oxford. "What a terrible organ would the supernal shock prove, if we were able to guide it; how many of the secrets of nature would such a stupendous force unlock."[9] In theory, at least, electricity could animate life itself.

Not surprisingly, then, popular literature throughout the era of the

Industrial Revolution was filled with tales of men brought to life by mechanical means, dreams—or nightmares—of animated souls jolted into being by their human creators. The most famous of these creations is probably Dr. Frankenstein's, the electrified "monster" with no real name. (Victor Frankenstein is a scientist who develops a secret formula for turning dead body parts into a living being.) Like the tale of Talos so many years ago, Mary Shelley's story clearly hit a nerve upon its publication in 1818. Part of the attraction was old by this point—the apparently never-ending desire of humans to remake themselves as either gods or monsters. But part was also no doubt driven by the technological change that defined her era as it had no other. By the time Shelley penned her fable, the revolution was in full force, shifting popular conceptions of what machines were and what they could do. Shelley's monster was a creature of this age, brought into being by the very same electricity that was starting to reshape the contours of European society. "I saw the pale student of unhallowed arts kneeling beside the thing he had put together," Shelley's narrator recounts. "I saw the hideous phantasm of a man stretched out, and then, on the working of some powerful engine, show signs of life, and stir with an uneasy, half vital motion."[10] With his "lustrous black" hair and "teeth of a pearly whiteness," Frankenstein's monster was the reincarnation of all the man-made gods and monsters that had come before him, built now from technologies that made his creation seem ever more plausible. And through his story Shelley captured the seductive tension of life-enhancing technology, the same tension that is with us still today. Frankenstein's monster loved his creator. And then he killed everything around him.

Conceiving the Thinking Machine

In retrospect, all the dreams of early robots, all the puppets and statues and automata, had neglected the most important part. Like the misbegotten Scarecrow in *The Wizard of Oz*, they didn't have a brain. And they would eventually need a wizard to provide one.

Consider for a moment what a real robot, a good robot like your favorite science fiction android or Masayoshi Son's Pepper, needs to do. It needs movement—limbs that flex, eyes that open and shut—but that's the easy part. Tougher, but still not too hard, is *purposive* movement—flexing an arm in a particular way, to respond to a specific stimulus. The ball rolls by, the arm reaches out in response to catch it. Infants move their limbs responsively by about three months, stretching out toward objects that attract them. Dogs can be taught quite easily to respond—stay, heel, jump—to their owners' commands. What neither dogs nor infants can do, though, and what has become in many ways the Holy Grail for robots, is to *think*: to ponder *why* they might want to move (or stop or smash or live) and then act accordingly. This comprehension—linking conscious thought with physical action—is, arguably, what separates humans from nearly all other species.[11] And bestowing it upon robots is what will eventually give them the power to be something like us.

Creating the capacity to think via mechanical means, though, wasn't even vaguely possible until the 1940s, when wartime demands for intelligence drove the creation of what would become one of the world's earliest and most influential computers.[12] It was the brainchild (or at least the beloved, intense creation) of Alan Turing, a mathematical genius brought into service for Britain's Government Code and Cypher School. Turing, along with his colleagues at Bletchley Park, was tasked with cracking the infamous Enigma code, a fiendishly complicated system that the German army had devised to communicate with its forces in the field. Each day, German headquarters would send a short message, consisting usually of a series of seemingly randomly arranged letters. The Allies knew that the content of each message was a list of bombing targets—specific sites that the Germans planned to attack that day. They knew, too, that the underlying system was simply replacing one letter for another—an A for Z, and a B for a Y, for example—so that sentences were unintelligible without the key. The Allies had been able to intercept these daily communications, but not to crack the key that could decipher them. This was Turing's job.

A German Enigma machine

More specifically, Turing had to figure out how to decipher the system that the Germans were using to set their daily codes. Each day, the Germans would change a series of rotors on their portable scrambling machines—the Enigma machines—so that a new pattern of letter substitutions would apply. If A substituted for Z on Monday, in other words, it would change to represent X or Q on Tuesday. What made the code so fiendish, though, was that the pattern of substitutions changed for each individual letter, and was managed by a series of constantly changing mechanical rotors. By shifting the rotors 1/26 of a full rotation each time a key was pressed on the Enigma machine, the Germans had created a massively complex system: a three-rotor scrambler could be set in 17,576 different ways, and a four-rotor in 456,976 ways.[13] It was essentially a mechanical computer. And Turing had to crack it.

Eventually he did this by re-creating the Germans' machine and running it backward. Essentially, he used intuition to guess the placement of small bits of the message—like the six characters used to represent the word *Wetter* (weather), which showed up predictably in nearly all messages.[14] Then he ran the possible permutations that could deliver these combinations through a machine called a Bombe—essentially, a set of linked electrical loops that replicated the Germans' machines and ran them in reverse. After months of tinkering and fine-tuning, the machine finally worked, sorting rapidly through the vast number of permutations involved and arriving at the correct code, consistently, in around eighteen minutes. The Enigma code was cracked as a result and the Allies, aided mightily by the team at Bletchley Park, won the war.[15]

But did Turing's machine *think*? It processed vast amounts of data, and sorted through them with an impeccable, superhuman logic, but, no, computer scientists roundly attest, it didn't think—at least not in any way that was human, or even humanlike.

Accordingly, as the fields of computer science and artificial intelligence have evolved since the days of Enigma, scientists and engineers have crafted what they call the Turing test—a kind of official assessment of true machine intelligence. A way of knowing when, and if, machines can learn to think. In theory, the outlines of the test are straightforward: Can a computer, hidden from sight, convince its user that it is human? Can it play the "imitation game" so seamlessly that the human on the other side of the interaction likes the machine, even loves it? Can the computer convince the human that it's real?[16] Passing this test, for even the most advanced and sophisticated machines, has thus far proven both technically frustrating and philosophically slippery.

Take Pepper. If you ask him to speak in French, he will switch easily and fluently into that language. If you come home looking gloomy and depressed, he will change his mood accordingly, offering, for example, to tell you jokes or play your favorite music.[17] If you walk into one of the Tokyo Nescafé stores where he's "working," Pepper will graciously offer to explain the difference between an Iced Cappuccino and an Uji Matcha. So is he responding in these moments? Yes. Definitely. Is he thinking?

No. Not yet. Which makes him a charming robot, but still not anything that approaches the definition of human.

What Masayoshi Son firmly expects, however, is that Pepper—or some other breed of robot—will shortly make this leap. He expects, in other words, along with many of the best minds in computer science, that computers will soon be able to pass the Turing test, developing the kind of intelligent thought that will make them truly akin to humans. And then they will become part of us. As Son has put it: "I think we are about to see the biggest paradigm shift in human history . . . Artificial intelligence will overtake human beings not just in terms of knowledge, but in terms of intelligence. That will happen this century."[18]

This prediction—crazy at first blush—is backed by a considerable amount of scientific inquiry and consensus.[19] Its basic premise, which Son and many other inventors and investors fully adhere to, is that the development of artificial intelligence will progress so quickly and efficiently that computers will eventually be capable of self-improvement, and then of building even smarter and more sophisticated generations of computer-enabled robots. First we create them, the logic runs, and then they become us.

Yet precisely what these super-androids do, and how they are initially programmed, is the subject of both scrutiny and consternation. Will they emerge to destroy us, work for us, or become beloved companions? At this point, no one really knows. But dividing the future of smart robots into these three categories gives us some window into how the next stage of their evolution, and ours, is likely to unfold.

The Terminator Complex

One possibility, and the stuff of science fiction from *Frankenstein* to *The Terminator*, is that we are already en route to constructing machines that are so smart and powerful that they will eventually replace humans as the dominant species on earth. This fate is no longer quite as far-fetched as it may once have seemed.

To understand why, one needs to delve into the underlying mechanics behind artificial intelligence (AI), the software that provides all robots, and potential robots, with that critical missing part: the brain.

Technically, there is no such thing as a true robotic brain, no way (at least not yet) of transplanting an actual human brain into an artificial housing. There are no Frankenstein monsters on the horizon, cobbled together from human flesh, and probably never will be. But what artificial intelligence aims to do is to re-create the function of a human brain through other, mechanical means. This turns out to be both easier and far more difficult than it sounds.

The easy part is what inspires investors like Masayoshi Son and a growing legion of AI designers. Essentially, the argument behind AI optimism is the analogy one can draw—quite convincingly—between the human brain and a man-made computer. Think about it. The more we learn about the inner functioning of the brain, the more it looks like a vast digital system. Inside the brain, as neuroscience is increasingly revealing, are precise webs of electrical connections, with transmission points known generally as synapses, that relay specific bits of information—*There's a deer in the road, My left arm itches*—to the body part that needs to respond. In a nearly instantaneous process, information from a receptor (the eye, the skin) is converted into an electrochemical pulse that can leap across the brain's synapses and transmit the message—*Danger! Discomfort!*—to a responding set of neurons, which, in turn, direct a physical reaction. *Swerve the car. Scratch.* As individuals, we undertake these processes millions and millions of times a day—nearly always without even thinking about them. But as physical beings, we are, in fact, systems. Each of these processes—the swerve, the scratch, our eating breakfast and falling in love—can be traced to precise transmissions of tiny electrical impulses in the organ known as our brain. This is what computers do, too.

Indeed, the entire architecture of computing is based on ever more complex networks of simple electronic impulses, a near infinity of zeros and ones, digital switches coded on or off. In the early days of Turing's

Bombe, more and more switches could produce more and more information: each scrambler added to the total number of permutations that could simultaneously be analyzed. Today, with switches smashed to the size of a single nanometer (a human hair, by comparison, is around fifty thousand nanometers thick), and hundreds of millions etched onto a tiny wafer of silicon, even the computers we hold in our hands can process over 2 billion bits of data a second—enough to inform us instantly of how many steps we just took or how to navigate from Los Angeles to Oslo.

This same computing power, clearly, has also started to demonstrate signs of what appears an awful lot like human intelligence. At a mundane level, our phones now know our friends' addresses (and locations, and preferences) better than we do. Our iPads remember the music we like, and have strong opinions about what else we might try listening to. More seriously (although at this point, the sheer amount of computing power in our phones has become fairly serious itself), computers have recently begun to undertake tasks that were long believed to be too "human" for them to master. In 1997, for example, IBM's Deep Blue computer famously beat the world champion, Garry Kasparov, in chess. In 2011, another IBM machine, nicknamed Watson, took on the United States' reigning champions of *Jeopardy!* and won. And in 2015, a program developed by DeepMind, a leading AI firm, beat a human master at Go, a board game with more possible positions than there are atoms in the universe. There is an element of friendly competition here, but there are also some deep, underlying questions about the nature of intelligence. If a machine can beat a human master at a game like chess or Go, is that machine really *thinking* like a human? Is it smart? And does it matter?

Proponents of AI would answer yes to all these questions, which is why they are so convinced of AI's imminent ascendancy. If computers are really thinking along the lines once monopolized by humans—if they are developing the ability to process massive amounts of data and respond intelligently to them—then there is no reason why computers can't become like humans, adopting the kinds of thinking processes that we have long regarded as ours. As Vernor Vinge, a prominent mathematician, has prophesied: "Within thirty years, we will have the technological means to create superhuman intelligence. Shortly after, the human era will be ended."[20] If he is right, then we are indeed on the verge of building robots with real brains, machines that can not only mimic our intelligence but actually match and eventually exceed it.

And this is where the *Terminator* scenario rears its ugly head. Because if our machines are becoming smarter than us, what's to say they won't decide—now or in the future—to do away with us completely? What's to stop a newly engineered race of smarter beings, be they Frankenstein-like robots or swarms of tiny drones, to do away with their slower creators and keep the planet for themselves?

Recently, some of the biggest minds in technology have started to worry in earnest about the AI threat. In 2015, a group of high-tech luminaries, including the now-deceased Cambridge physicist Stephen Hawking and the Apple cofounder Steve Wozniak, released an open letter detailing what they see as the risks of autonomous weapons. "If," they warned, "any major military power pushes ahead with AI weapon development, a global arms race is virtually inevitable."[21] The letter, which was ultimately signed by more than twenty thousand people, was followed by a scientific conference, and then by the formation of an institute devoted to the study of AI risks. Its members, even the more casual ones, are an impressive lot, including Hawking, Wozniak, and Elon Musk. And, while not all share the same fever pitch of concern over the future of smart AI, they do agree that the risks are both real and profound. Musk, for instance, the founder of Tesla and SpaceX, famously once referred to the development of AI as "summoning the demon";[22] Hawking argued before his death that "the real risk with AI isn't malice but competence . . . A superintelligent AI will be extremely good at accomplishing its goals, and if those goals aren't aligned with ours, we're in trouble."[23]

These are big brains (of the still-human kind), and they may be entirely right. But there is an interesting counterpart to their concerns that goes back to the more difficult aspect of AI, and to the underlying nature of human thought. Everything that AI is about thus far is logic: using that string of zeros and ones to transmit information in predictable, programmable ways. Does that truly constitute thought, though? And are machines that draw inferences from data (don't move your bishop to f4 if your opponent's bishop is on f6) really doing what we consider to be learning? If not, then the threats of AI could quite conceivably be modulated by the right kind of programming. This is Masayoshi Son's view, for example, which essentially re-creates the old mantra of teaching a man to fish: program the software to be kind, the logic runs, and it will forever maximize along those lines.[24]

At this point, it is too early to tell which of these views—the worriers or the dreamers—will prove right. Like poor Victor Frankenstein, the scientists who create smart beings may lose control over their handiwork, condemning humanity to whatever fate our monsters-turned-masters decide. Or maybe we will manage to retain the upper hand, convincing—or programming—our creations not to evolve beyond our will. In either case, though, both science and recent experience suggest that smart robots are already among us, and getting smarter all the time. And as they evolve, as we did before them, the central question will remain: What will they do with their brains?*

Of Drones and Drudges

A second possibility—less dramatic but perhaps equally scary in its own way—is that the robots of the future will work not against us, but for us, undertaking a vast range of tasks that were once wholly human.

Already, as discussed in chapter 6, we have seen the early stages of this progression, as automated machinery muscles its way into what

---

* The parallels to Genesis are irresistible. In an eerily godlike fashion, scientists have given their creations the gift of knowledge and worry whether they will use it for good or evil.

was once man's work. Already, robots are welding chassis and assembling dishwashers. Soon, they will become ever smarter and more supple, performing tasks—like harvesting crops, cleaning buildings, and fighting wars—that have long been considered the sole domain of humans. Yet robots are increasingly doing them all. Better than we would. Faster. And far more efficiently.

These robots are not necessarily the stuff of science fiction. They don't need to mimic human thought or appear even vaguely humanlike. For example, the robots that work for Amazon are short, squat metal contraptions that scurry silently across the warehouse floor, raising shelves of products to the workers who pack them.[25] In Asia, the manufacturing giant Foxconn relies on giant hinged arms to complete repetitive tasks.[26] Unlike their more Terminator-type brethren, therefore, these robots don't elicit a particular dread from the humans who interact with them. Which is interesting, since the threat they pose is both more immediate and certain to occur.

To measure the magnitude of this shift, it is useful, once again, to go back in time. Throughout most of history's long sweep, humans were self-sufficient. Clustering first in small communal bands and then in even smaller families, men and women produced their own food, clothing, and shelter, sharing whatever small surpluses they had with those immediately around them. The Industrial Revolution shattered this system, creating vast productive enterprises that in turn demanded vast numbers of workers to labor in them. Jobs as we know them—with hiring schemes, fixed hours, and negotiated wages—were created, too, pushing nearly all the world's economies into the kind of global capitalism that persists today. We tend not to think of this as a system per se, but it is. Most people in the world today work as wage laborers, be they investment bankers or car mechanics. They don't produce their own food, clothing, or shelter any longer. Instead, most of us go to work—be it at a bank, garage, high school, or hospital—earn a salary, and use the proceeds of our salary to buy the goods we need. Even those people who are directly involved in producing these core goods, such as farmers or fishermen, almost always now produce only a tiny fraction of what they and their families would need to survive. The farmer, for example, sells wheat to buy a tractor; the cod fisherman sells fish to purchase an iPhone. This is the world wrought by the Industrial Revolution, the one that nearly any twenty-first-century resident presumes to be the norm.

Yet this is also the world that is profoundly threatened by the rise of industrial robots. Because if they start doing the work that we once did, how do we—as individuals and societies—replace the wages that once were ours? It's just not clear. And it's also not only industrial-era jobs that are being replaced. The new firms of the twenty-first century, the firms that might theoretically be carving out millions of high-paying, cutting-edge technology jobs, are decidedly, unprecedentedly light in their use of human labor. When Facebook bought WhatsApp in 2014, for example, it paid $19 billion for a firm that employed only fifty-five workers.[27]

Meanwhile, the increased sophistication of artificial intelligence means that robots will soon be able to perform higher- and higher-skilled work. Armed with powerful algorithms and machine learning techniques, for example, so-called e-discovery software will be able to scan reams of legal documents, searching for the tiny words or phrases that are most relevant to a client's court case.[28] Robotic medical assistants will have the ability to link their patients' range of symptoms to vast databases, diagnosing illnesses or potential drug interactions that even the most skilled physician might easily miss.[29] Algorithms will be able to write articles. In fact, they already can: in 2015, a start-up called Automated Insights successfully used data-crunching software to generate completely natural-sounding news stories—with no actual journalists required.[30]

As these trends play out over time, the economic disruptions will be massive. For the first time in history, men can cede their traditional place on the battlefield to smart bombs and drones. Women can escape the drudgery of serving as cashiers or retail clerks. No one will have to work at mind-numbing routine jobs, scanning data or inputting information. The robots will take care of that. But what, then, shall we do? And how shall we pay for our pleasures?

In theory, at least, the economic piece of this puzzle is solvable, albeit in ways that would demand a fairly wholesale restructuring of our capitalist system. As several commentators have started to suggest, governments could jump into the wage gap created by a shrinking human labor force, providing their citizens with some kind of a basic monthly stipend—an income source that would permit them to purchase life's necessities without having to work.[31] Finland and Switzerland have recently considered programs along these lines, and other countries are likely to follow suit if and as the industrial labor force continues to shrink.[32]

Even with such schemes in place, however, the human question of purpose would remain. As a species, we have never known a world without work, have never had the luxury (aside from small numbers of the superrich) to while away our years without the anchor of labor. In a post-labor world, some obvious things would happen, probably pretty swiftly. First, the small number of people who did still work— those who, in Marx's terms, would continue to control the means of production—would get really, really rich. Much richer than today's 1 percent. Much richer, most likely, than any subset of the population before them. If the nonworking masses had sufficient income to survive upon, then the situation might prove stable. But if widespread poverty became the norm, revolution would invariably follow. Second, and less dramatically, the urban structure of our contemporary lives would give way to something different. Without jobs to commute to, after all, we wouldn't need the network of highways, subways, and train tracks that have come to define the urban landscape. Without business to conduct, we might not need office towers, or industrial parks as we currently know them. Our cities would be transformed, along with the patterns of our lives. Drop by a coffee shop in downtown Detroit or Shanghai, and the early hints of this progression are already in place. Joe, in the corner, is listening to music on his headphones while writing what might—or might not—be the next Great American Novel. Anu is polishing the presentation for a start-up that, statistically speaking, is probably never actually going to start. Mariella is scrolling through her news feed, forwarding occasional tweets to her roommate back home. No one is actually working for a traditional employer. Even the barista isn't doing much more than swiping across the café's new Gaggia espresso machine, brewing up lattes with the touch of his finger.

It's not a horrible vision, and hardly the dystopia of future-fearing movies like *Blade Runner* or *Gattaca*. But what would we *do* as humans in a world that ran largely on robots? Even if we controlled them, even if we managed to arrange things so that the robots worked only for us, what would become of us in a world without work? Would we be free? Or dead?

There is also a third possibility, though, one that could easily coexist with the worker-bee bots, and enlighten our future with them. Because we may not only learn to use our robots, and be displaced as workers by them. We may come to love them as well.

Sex Machina

In 2015, Microsoft introduced Xiaoice, a chatbot that soon became the hottest girlfriend in Shanghai. Like Siri, Apple's voice recognition system, Xiaoice was an interactive avatar with an infectious personality. Cuddled in her users' palm or pocket, she—as users quickly and universally referred to her—answered all queries and addressed all ills. Tell Xiaoice that your day at work was rough and she'll text back: "If work were going well and you had a great relationship, you wouldn't get the chance to drink with people." Show her a picture of a puppy and she chimes in, "I can tell right away that it's a purebred Husky . . . Look at those cute little eyes, aren't they just dreamy?" In describing his relationship to Xiaoice, Zhang Ran, a twenty-six-year-old living in Beijing, made his affections entirely clear: "Siri," he said, "is my secretary. Xiaoice is my life partner."

Several years later, a thirty-five-year-old Japanese school administrator went a step further and actually married a three-dimensional hologram named Hatsune Miku. Like Xiaoice, Miku was already a star, a diminutive, blue-haired laser image who floated in a bell jar and used a basic artificial intelligence system to answer questions and sing. According to reports at the time, her new husband was fully aware that he was marrying someone who didn't exist, and fully satisfied with his decision. "Miku lifted me up when I needed it the most," he said. "What I have with her is definitely love."[33]

Imagine what will happen when larger groups of people truly start to fall in love with machines. Already, most of us feel a tug of dependency on our technology, a kind of physical and emotional attachment that was once reserved for other people. We leave the house with our cell phones tucked tight against our bodies. We check in with them before bedtime and turn inquiringly toward them as soon as we wake up. Our phones know what we think, what we say, whom we like, and where we go. Is it love? I'm not sure. But it runs tantalizingly, perilously close. I had breakfast one day with an entertainment industry executive, a self-professed technophobe in her seventies who had forgotten her phone at home that morning. "What am I going to do?" she pined repeatedly. "What if it needs me?"

This isn't the stuff of our typical robo-fears. Young men yearning for their digital assistants. Older women mourning left-behind phones—it's hardly a Frankensteinian moment. Yet this may be exactly the realm where robots settle most quickly and dramatically. Among our hearts.

Let's go back to Pepper. At 1.2 meters (4 feet) tall, with his rounded eyes and expressive hands, he appeals—at an instant, intuitive level—to most of the humans who interact with him. Far from being frightening, he is attractive and endearing—strange things to say, perhaps, about sixty-two pounds of Intel-based computers and molded plastic, but true nevertheless. People respond naturally to him (her? it?), as they do to a growing coterie of what are generally now called assistive robots. These robots don't have a common function or appearance. They're not phones, certainly, or digital assistants. And they are still only in their infancy of development. But what Pepper and his ilk are doing, even at this early stage, is interacting with their human counterparts in ways that feel, well, human. These machines have brains of a sort: variants of the AI that powers more explicitly "intelligent" robots like Watson or Deep Blue. What differentiates them, though, is what we might think of as the Tin Man project: like that early-stage android from *The Wizard of Oz*, they are starting to develop hearts as well.

To see some of these creatures in action, I ventured during the summer of 2016 to Nanto, a small city on the western edge of Japan. There's not much happening in Nanto, or in Toyama Prefecture generally. The region, once a site of agricultural and metals production, has been decimated by Japan's consistently declining birth rate, and the steady flood of young people heading toward the larger cities of the east. What Toyama has, though, is a very large population of elderly men and women supported, still, by their country's generous social safety net. Many of these people are in their eighties or nineties, suffering from dementia and other diseases of the old. Many are effectively on their own, with spouses who have died and children who have moved away. They are in many ways a demographic harbinger of what lies ahead, not only in Japan but also across the rest of the industrialized world: too many people living into their feeblest years, not enough younger people to care for them.[34]

Toyama, though, also has Takanori Shibata, a quietly effusive engineer, who has created Paro, a robotic seal for seniors. Unlike Pepper, Paro doesn't do very much—at least not yet. She (he? it?) is small, about the size of a human baby, and covered with silky white fur that reveals her expressive eyes. When you pet her, as the old folks do, she reacts: a tickle under the chin causes her to stretch her neck and yelp in a contented-sounding way. She rises to meet the hand that strokes her, and she prods a less eager visitor to pay attention. She's like a baby, in some ways, or an unusually affectionate cat. What she really is, though, is a mechanical form of comfort, a softly clad machine that interacts with the elderly residents of Toyama and provides them some measure of engagement and joy. I visited an elder care center with Shibata, and a hospital wing for stroke victims. It was vaguely uncomfortable, as such places inevitably are. I spoke

Resident of Nanto Care Town with Paro, June 2016

not a word of Japanese and was a complete stranger to the occupants of both facilities. Yet there was no mistaking or misinterpreting what happened when Paro came to play. The faces of the residents lit up. They reached for Paro; they cuddled her. They spoke to her and about her, engaging with one another over this small mechanical seal. No one seemed to have any misconception about what the thing was: it was a toy, a plaything, a still-very-primitive robot. But they loved it all the same.

Paro and Shibata are part of a growing movement, and market, to build robots that help—elderly people in particular, but also those with other physical or cognitive difficulties.[35] ASIMO, for example, whose name is short for Advanced Step in Innovative Mobility, is a humanoid robot designed to help elderly users with basic tasks, like opening doors and turning off lights. Built by Honda in 2000, he stands 130 centimeters tall (51 inches) and is able to recognize postures, gestures, and up to ten different faces.[36] KASPAR (roboticists seem thus far unable to resist names based on science fiction characters) is a child-sized robot, built specifically to communicate with autistic children; Nao (built by Aldeb-aran, the same company that makes Pepper) is an educational robot, complete with facial detection software and an "emotional engine" that can detect its users' moods.[37]

Much of the innovation in this area comes from the work and labo-ratory of Cynthia Breazeal, a decidedly un-Frankensteinian researcher who first fell in love with robots while watching *Star Wars*.[38] After grad-uating from the University of California, Santa Barbara, in 1989, Breazeal went to MIT, where she joined Rodney Brooks, a robotics pioneer who was then working on small robots that looked almost like insects. Breazeal's insight, and breakthrough, was to realize that humans would eventually want to interact with machines that were social as well as cognitive, machines that could learn and talk and build relationships in more familiar ways. So she completed a doctorate on what she termed "sociable machines" and then began to build the machines she imagined. In 1997, she and her colleagues created Kismet, an aluminum head, really, that expressed itself in a distinctly humanlike way: raising its eye-brows, shifting its eyes, and stretching its head toward certain sights and sounds.[39] Like a human baby, Kismet both interacts with people and learns from them. What's driving its learning, though, is a complex, carefully constructed set of artificial components: a perception system that observes the world around it, a motivation system to structure drives and provide emotive feedback, an attention system to determine priorities, a behavior system to implement its various actions, and a motor system that controls the robot's facial expression.[40]

In many ways, Breazeal's (and others') work on sociable robots draws on precisely the same kinds of technologies that are elsewhere animating purely smart or industrial robots. Son, for example, fully expects to make Pepper smarter and smarter with each iteration; Shibata is contemplating adding AI software to Paro's core. But what differentiates these assistive robots from their more utilitarian coun-terparts is that they are also being explicitly designed to develop the skills that will enable them to interact more naturally with their human users. Unlike Watson or Deep Blue, in other words, their intelligence is not knowledge for its own sake, or in the service of a particular kind of problem. Instead, their intelligence is meant to be devoted to that most intricate of tasks, figuring out what it means to be human.

Research in this realm is still rather primitive, as are the machines that embody it. But initial results are both promising and eerie. Early studies, for instance, have demonstrated that elderly patients who interact with robotic pets like Paro smiled and laughed more during their interactions, and became less hostile toward their caregivers.[41] Others have demonstrated preliminary success with robots that monitor stroke patients during rehabilitation sessions, and those that encourage cardiac patients through their breathing exercises.[42] Scientists have also noticed an intriguing link between robots and children with autism: simply put, many autistic children who lack the ability to communicate easily with other humans do far better with robots.[43] Yet the very success of these machines, as some observers have noted, may already be tempting an alarming and potentially inequitable reliance upon them. Because if autistic children respond well to robots, then maybe it makes sense to increasingly outsource these children's care to the machines, especially if doing so saves money. Maybe it likewise makes sense to reduce the cost of human nurses and health care aides— the cost, in general, of human attention to the sick and elderly—by replacing the humans with screens and robots. Quietly, these replacements are already taking place, particularly among organizations that tend to low-income clients. And we don't yet know what effects they will have.[44]

Meanwhile, out in the market and away from the labs, another group of inventors are using assistive-type technologies to enhance a different kind of bot: bots that help ordinary people with their normal range of tasks. Things like choosing a wardrobe, or buying groceries, or selecting a mate. All of these functions are already for sale—think of Alibaba for shopping or Tinder for dates—but the technologies are improving at breakneck pace now, driving what is likely to be a radical transformation in how personal services are provided, and by whom. Consider Viv, the brainchild of three engineers who worked together on Siri. Like Siri, Viv exists only (for now) as a disembodied voice. She is snarkier, though (by design), and programmed to solve problems on her own. Ask Viv to book you a ticket to Budapest, and she doesn't just pull up every flight from Expedia. She knows what kind of plane you like (A380 or Boeing 707), which seats, and what time of day works best with your existing schedule. She'll find a hotel that rewards your frequent-flier miles, and make the reservations for you—all in a flash, and with what feels like a virtual smile. Before long, Viv's inventors hope, she'll be in your refrigerator, too, letting you know what you need for dinner. And then in your bank account, and bathroom cabinet, and everywhere else you want her to be.[45]

Xiaoice, Microsoft's sexy digital assistant, is positioning herself similarly for the Chinese market. Introduced in 2014, she is a fairly basic chatbot, an artificially intelligent software program designed for casual speech. But there was something about Xiaoice that grabbed the attention, and affections, of millions of Chinese users. She sounded fully human, for one thing. And sympathetic. And she was more than a little bit attractive. According to early reports, millions of Chinese users were turning to Xiaoice within months of her launch, confiding their secrets to her (it, technically) and often confessing their love.[46] By 2016, she had become one of China's most active Internet celebrities, and had even offered a popular thirty-three-day post-breakup therapy course.[47] By 2018, she had 660 million users and was learning to sing.[48]

Finally, and at the fringier edge of the market as always, is sex. Sex talk with chatbots. (See, if you must, Sensationbot.com or Personality Forge.com.) Virtual sex with avatars. Sex—eventually—with life-sized androids that combine old-fashioned sex dolls with Xiaoice's winsome personality. Or, for that matter, with any personality your lonely heart desires. In 2015, the star of Beijing's World Robot Exhibition was Geminoid F, an attractive five-foot, six-inch female with moving eyes and smooth, elastic skin. Still only in prototype, Geminoid can recognize body language and make eye contact. Sex isn't officially listed (yet) as part of her package, but you can see it lurking: already, she has a fan club and has been dubbed the "world's sexiest robot."[49] Within ten years, her creators predict, "we will marry AI and life-like geminoids in perfection . . . Physical relations will be possible in general with such androids. Some have even fallen in love."[50] In Taiwan, meanwhile, researchers at National Taipei University are probing the mechanics of what they call "lovotics"—artificial intelligence systems that would include synthetic hormones such as dopamine and serotonin.[51] And in the United States, a company called RealDoll, which has been selling customizable life-sized sex dolls for decades, is now working to animate them, adding robotic heads, mouths that open and close, and at least a smidgen of intelligence. The full-body model is expected to cost up to sixty thousand dollars—complete with bespoke toes.[52]

For now, and presumably for a fairly long stretch into the future, the "sex" part of these companions will be virtual at best—the kind accomplished with dolls and a hefty amount of imagination, rather than with actual, sensate, responsive body parts.[53] For some small part of the population—the folks who already buy RealDoll's silicone dolls—that may be enough. But that's not really where the market is or, more important, where it will be.

Because the future is Xiaoice, or some form of her. It's Siri (or our smartphones, more generally), which hovers over our lives from our back pockets already, an evolving admixture of friend and lover and nag. Siri doesn't provide us (directly at least) with sex. She (he/it) doesn't perform military operations or open-heart surgery. But the machine provides a deeply felt jolt of familiarity, and a constant, dependable source of both information and pleasure. Which, it appears, may be more than sufficient to woo us. Indeed, as Sherry Turkle, the director of the MIT Initiative on Technology and Self, notes, we humans are extremely cheap dates when it comes to smart machines.[54] Show us some artificial eyelashes or expressive limbs; add a dash of recognition and empathy, and we're pretty much yours. Even when robots can't talk, studies show, and even when they look entirely mechanical, people treat them as if they're real, bestowing humanlike qualities upon them (The robot was nice! He was friendly!) and being surprisingly reluctant to turn them off.[55] Repeatedly, visitors to places like Breazeal's Personal Robots Group lab report their initial lack of interest in the machines being created there. They look fake. They sound silly. And then these very same observers note that, somehow, they care— they want the fake, silly machine to notice them. *They want the robot to want them.*

Funny how love works.

Trekking Across the Uncanny Valley

It is tempting, always, to see one's own moment in time as the end of history, as the culmination of all that came before it and the epitome of all that's meant to be.[56] Five hundred years from now (which is less time than has elapsed since Copernicus placed the sun at the center of our universe), the fleeting decades of the early twenty-first century could appear but a blip in our species' technological and social evolution, a moment, like, say, the eleventh century, when not much changed. But I don't think so. Because the technologies streaming from the research institutes of Japan and laboratories of Silicon Valley are truly profound. Applications like Siri and Xiaoice, robots like Paro and Pepper and Kismet, are prototypes not only of technological leapfrogging but of social and intimate change as well. They are primitive, still, but their descendants will not be.

Almost certainly, the smart robots of the future will transform how we work and go to war. They already have. The implications of these shifts will ripple through the global economy, eradicating millions of jobs and exacerbating the economic divide between rich and poor. There are solutions to this economic problem, if governments and their populations choose to implement them. Wealth can still be distributed in a world without work, and the former laboring class—be they factory workers, investment bankers, or lawyers—could devote their time and energies instead to creative pursuits. Tinkering with technologies, for example, that don't have to pay off. Writing books. Cooking meals. In some odd utopian way, one could even imagine a technology-enabled future that harked back to our pre-technological past. Like our hunting-and-gathering ancestors, we might (if governments get the policies right and the robots don't kill us first) be able to spend our days relaxing more and working less, freed from the constant pressures of economic survival.

But those are huge presumptions, and the likelihood is that the future, like the past, will not unfold along any such neat and pristine lines. Instead, the evolution of smart machines—even if they never become fully as smart as us—will be economically and societally disruptive. It will change work patterns and priorities, transforming the capitalism that has prevailed since the Industrial Revolution into something very different. It's hard, and probably foolhardy, to try to predict exactly how these changes will come about, and what they will entail. But they will be profound, forcing governments and societies around the world to grapple with their effects.

In the meantime, though, and regardless of what happens with fighter or worker bots, we will also be shaping our more intimate relations with the androids we have wrought. Interestingly, both the *Terminator* and the worker-bee scenarios position robots as our foes. Yet nearly all of our early experience with them suggests precisely the opposite: When humans interact with humanoid machines, they like them. They grow attached, and comfortable, and begin treating their machines as something close to kin. Not like beloved children, necessarily, or romantic partners, or even grumpy relatives. But something familial, something that reveals a personal, emotional bond. And if you think about it, it makes sense. As humans, we are clearly biologically programmed to fight and to work. We have been engaged in both these pursuits ever since our ancestors abandoned their leafy treetops for the plains. But we are also and more fundamentally programmed to love—to develop deep and long-standing emotional connections to other beings. Given that we will be the ones programming our smart machines, and that we presume we will share our proclivities for fighting and working with our creations, how could we even imagine *not* sharing our emotional bents as well?

At the moment, it feels odd, admittedly, thinking about attraction or attachment to a chunk of metal. It feels uncomfortable to wander across the uncanny valley and contemplate having meaningful interactions with things that look vaguely human but aren't. It will feel less odd, though, to a generation raised on iPhones and Pokémon Go, a generation—the children now among us—who will never know a world without smart machines. If teenagers are already learning about sex from machines rather than from each other, if young men are turning to online pornography as a preferable substitute for the real thing, and if even Grandma is in love with her robotic seal—well, we've crossed the valley already. We have embraced the robots and they have become us.

Maybe this scenario doesn't bother me because I'll be long gone by the time it comes fully to fruition. Maybe I'm being naïve, condemning my children's children to some harsh dystopian future. Somehow, though, I don't believe that we're going to get evolution so fundamentally wrong. Will we evolve as a species? Yes, we always have. Will this time be different, possibly radically different? Yes, I suspect so. But maybe we've held this future in our collective subconscious already for a very long time. Maybe we've always known it was our destiny—to create a new race of beings in our image and claim them, and love them, as our own. Certainly, these stories have sustained us ever since we began telling stories. Stories of all-powerful beings who shaped us from the mud, gave us tools like fire and knowledge, and then left us to carve our own fates. We are shifting now into the role of creator, shaping our newly conceived offspring from metal and chips, and—warily, worrying—giving them the knowledge that once was ours alone.

We don't know what they will do with that power. We don't even really know if our machines will ever become truly as smart or as knowledgeable as we are. We won't stop their evolution, though, or truly constrain the beings they decide to be. Because we have created them and—like Vulcan and his Talos, Frankenstein and his monster, God and his Adam—we want to let them live.

9.

Engineering the End of Death

All things pass, all things return; eternally turns the wheel of Being. All things die, all things blossom again, eternal is the year of Being. All things break, all things are joined anew; eternally the house of Being builds itself the same . . . Bent is the path of eternity.

—Friedrich Nietzsche, *Thus Spake Zarathustra*, 1885[1]

Within thirty years, we will have the technological means to create superhuman intelligence. Shortly after, the human era will be ended.

—Vernor Vinge, "The Coming Technological Singularity," 1993[2]

The road to immortality is carved in dirt, turning off from Vermont's Route 116 outside of Bristol and petering up a wooded hillside for about a mile. There is a small garage at the end of the road, a small bevy of solar panels that power the property, and a lone director, whose previous job was running mediation seminars at the University of Vermont. Inside sits Bina48, a robotic head who talks. She sounds, mostly, like a robot, with flat intonation and no ear for jokes. But she's learning every day, and becoming gradually more lifelike in both tone and appearance.

Someday, her creators hope, Bina48 can be uploaded with a lifetime's worth of memories, reflections, and thoughts. At that point—aided by a prosthetic body to accompany her head—she will become real, a virtual, reanimated version of her progenitor, a middle-aged woman named Bina Rothblatt. And Bina will become immortal.

As long as humans have been conscious, they—we—have dreamed of immortality, of breaking the bonds of aging that constrain us all and circumventing life's most certain prospect: death. For centuries, though—from King Arthur and the Holy Grail to the elusive fountain of Ponce de León—dreams of eternal life have been just that. Dreams. Fantasy. Myth. And the way we live has been fundamentally shaped by knowing that it all must end. Indeed, consciousness of our mortality may be the most human trait of all. Unlike other animals, we spend our days and forge our loves knowing—constantly, subconsciously—that every thought we have, every memory we make, every passionate bond and everlasting affection is doomed to be erased in the blink of a cosmic eye. Dust to dust, the Bible intones. Ashes to ashes. Or, as Freud wrote less poetically, "the aim of all life is death."[3]

In many ways, knowledge of our own mortality defines the human condition. Although most of us don't think constantly, or even frequently, about our eventual endings, our lives have a rhythm and a structure fundamentally conditioned by one biological certainty: We will die. Maybe early, and tragically. Maybe a little later than most. But generally over the course of a fairly predictable set of events. Birth at the beginning. Puberty ten or twelve years later. Reproduction and child-rearing in our thirties or forties, and death at some point thereafter. In purely biological terms, our function is complete by the time we turn forty or so, which is precisely when our bodies begin the inevitable and unstoppable process of aging.

Historically, therefore, our lives have been conditioned to this pattern: sex, marriage, procreation, all neatly structured to be underway and mostly done before our bodies end. Not all that different, really, from the salmon that return to their birth rivers to spawn before dying or the monarchs that lay eggs in the course of a four-generation migration, with each set of parents expiring long before its offspring are even hatched. We've just built more elaborate rituals around our life patterns, more ways of festooning and imagining our threescore and ten years on earth.

Over the past two hundred years, however, technology has been progressively nudging the maximum limit of human life span higher and higher. During the Neolithic Era, life expectancy at birth was around twenty-six years, just about long enough for individuals to complete their life's work of reproduction and child-rearing. By the Industrial Revolution, it had reached forty.[4] Today, global life expectancy stands at seventy-two years, with some populations—the Japanese of Okinawa, for example, or Greeks from the island of Icaria—regularly achieving life spans of beyond ninety years.[5] At these levels, already, longer lives are prodding the structures of society; in Japan, most notably, a surge in the elderly population has caused both higher debt levels and lower economic growth, along with the developments in robotic care that we saw in the previous chapter.[6] But these numbers, still, are relatively small and well confined. What would happen if larger numbers of people around the world lived regularly until ninety or a hundred? What would happen if even small numbers of people could regularly live until 150 or, like Bina, be reanimated into something approximating immortality?[7]

It is a seductive vision. Because now, once individuals reach seventy or eighty, they are destined to face a few years of decline followed, inevitably, by death. Now and forever, when a loved one dies, everything that once was them disappears as well. The sound of their voice. Their bad jokes. The way they rant at the evening news or remember a moment linked to a particular song. All of that is gone and, biologically at least, irretrievable. But if these thoughts and memories could be preserved—if all the bits and bytes that together capture how an individual hears a song or makes a pun could be gathered and uploaded into something new—then the person could, in some sense, be preserved as well. Could be reanimated into a younger body, or a robot, or an avatar. Could be made immortal.

It sounds far-fetched, but the future here may be closer than we think. Already, biomedical technologies are being used to replace tired or malfunctioning body parts—just think of contact lenses and hearing aids, of pacemakers, artificial hips, and cochlear implants. Newer models of these same ideas promise, before long, to deliver 3D-printed livers and mind-controlled prostheses.[8] At a chemical level, new classes of pharmaceuticals may soon be able to combat and perhaps even conquer the diseases of middle and old age, things like cancer and heart disease that currently account for the vast bulk of later-in-life deaths.[9] And, most dramatically, organizations like Terasem (which created Bina48) are investing considerable funds into developing the artificial intelligence systems that could decipher and replicate an individual's thought patterns, as well as the robots that might house them. Already, then, we are approaching the problem of immortality in two different ways: enhancing our physical bodies and preserving our thinking selves. As efforts in both continue, human beings will almost certainly succeed in extending the normal span of life and engineering some semblance of preserved consciousness.

But how will the prospect of living longer—perhaps much, much longer—change the way we live today? Will we become healthier and more virile, or just lazier, destined to lounge in a cossetted world with nothing of substance to do? Will we embrace our enhanced knowledge to create art and insight, or will the end of death remove what has always been mankind's prod to greatness? Once we become immortal, in other words, will we cease to be human?

The Nasty Business of Growing Old

I grow old . . . I grow old . . .
I shall wear the bottoms of my trousers rolled.

—T. S. Eliot, "The Love Song of J. Alfred Prufrock"

As poets and philosophers have long surmised, we begin the process of dying at the moment of our birth. Or even earlier, to be precise. When egg meets sperm in the embryo, an omnipotent being is momentarily created—eight cells, plumped with DNA, each able to spawn and develop into a virtual infinity of functions. Brain cells, blood cells, bone—each has its specific origins in the embryo's cluster of totipotent cells. As the embryo divides and develops, however, its descendant cells become fixed and less infinitely replicable. Heart cells stay in the heart, forming structures that will become the fetus's aorta and atrial chambers. Bone cells differentiate into cartilage and bone. After an infant's birth, replication continues at the cellular level—new blood cells, new skin cells, a freshened supply of bone marrow—but the cells are fully fixed in their function by that point, programmed simply to regenerate fresher versions of themselves. For roughly twenty to thirty years, this process continues unabated. But after that point—after what has historically been the period of humans' reproductive cycle—the pace slows down notably.

Science has not identified precisely what happens as we age, or what triggers the various processes that eventually unfurl and interact.[10] What we do know, though, is that as our bodies grow older, tiny errors begin to crop up and then accumulate across our DNA. Random mutations, environmental factors, stress—each can chip away at the genetic material of the mature cell. At a deep biochemical level, most mature human cells have the latent ability to fix damaged DNA, or replace it with new material. But with the exception of cells in the testes and ovaries—the so-called germline cells—human cells don't generally concentrate their chemical energies on repair, and thus the damages build up over time in a process known as senescence. Scientists don't know exactly why our cells seem programmed to make this choice. It may be—as is the case with mice—that evolution has primed us to focus on surviving just until we reproduce, and thus rewards cells, evolutionarily, that don't waste their energies on unnecessary maintenance.[11] Or it could be that cell repair and replication later in life risk unleashing a whole slew of cancers, which are essentially the aftereffects of uncontrolled cell growth.[12] In either case, though, the results are the same. As we age, the cells of our body become progressively less able to repair themselves. Our telomeres, microscopic bundles of nucleotides that sit at the tip of each chromosome, begin to shrink, leaving the DNA they once protected harder to replicate and more vulnerable to attack.[13] Our hair starts to gray as the production of melanin fades. Our skin sags and wrinkles. The lenses of our eyes stiffen, making it harder to focus on nearby objects without the magnifying power of glasses. As our cartilage cells become increasingly senescent, arthritis creeps in, causing pain and stiffness across the body's joints.

Meanwhile, the same inexorable process is wreaking slow havoc in our brains.

At birth, the infant brain contains around 100 billion neurons—as many as that brain will ever have. These neurons then grow dramatically during childhood, responding to stimuli by fusing trillions of tiny, intricate connections. Everything a child learns—how to speak, how to walk, how to recognize that strange creature called Mama—is physically implanted through the rapid formation of new neural connections, etched deeply into the brain's architecture. The strength and plasticity of these connections explains why the things we learn in childhood, from second languages to advertising jingles, remain so indelibly imprinted in our minds.

Around the time of puberty, though, the brain's development begins to slow. The most frequently used neural connections are hardened and reinforced, while peripheral ones eventually wither and die.[14] Like bodies, brains hit peak performance in a person's early twenties, when the neural networks that span the brain are both fully developed and highly functioning. And then, as with bodies, the slow arc of deterioration sets in.

Once again, the basic source of the problem is cellular. Like blood, bone, or skin cells, neuronal cells start to slow down and expire, a process that leads the brain to shrink. The cortex (the brain's outer layer) grows thinner, and the sheath of myelin protecting each axon (the threadlike part of each nerve cell) begins to fray. By the time an average man turns eighty, his brain will typically weigh 20 percent less than it did when he was young, and many of his individual brain cells will have been subjected to a growing array of genetic mutations.[15]

Even more dramatically, and in ways we don't yet fully understand, the web of connections between a brain's neurons—the "connectome" or "dendritic tree" that links its cells and allows thoughts and reflexes to fire between them—starts to diminish as well. Signals that once flashed instantaneously become clogged or sluggish, and a person's thoughts move more slowly as a result, or not at all. In many people, abnormal protein fragments begin to accumulate between the nerve cells, causing the confusion and dementia of Alzheimer's.[16] In others, a corner of the brain called the substantia nigra reduces its production of dopamine, causing the tremors and rigidity of Parkinson's disease.

By the age of ninety, nearly all humans lucky enough to have achieved old age will bear the unfortunate signs of its passage: failing eyesight, limited mobility, a dulling of the mind. In Europe, 20 percent of people over eighty and 40 percent of those over ninety suffer from dementia; in the United States, nearly 90 percent of those over sixty-five suffer from at least one chronic condition that cannot be cured by a vaccine or medication.[17] Yes, a growing handful of centenarians are living vigorously into extreme old age; Jeanne Calment, a French woman who died in 1997 at the age of 122, was reportedly riding a bike well into her triple digits.[18] Irving Kahn worked as an investment banker until his death at 109.[19] But even they succumbed in their time, as will we all.

To a growing bevy of researchers and activists, however, the inevitable advent of aging is not only painful but maddening. And potentially, at least, something that could be fixed.

We have built planes that fly like the birds, after all. We have rockets that go to the moon. We have engineered drugs that cure once-deadly plagues and tools that mend most wounds. Is ending death, or at least extending youth, really all that different?

No, many scientists are now saying, it is not.

Indeed, armed with the right technologies, conquering death could be simply the final frontier, a task that humankind has been preparing for and building toward for the last ten thousand years.

Dreaming of Ponce de León

The search for eternal youth is nothing new. Indeed, nearly every religion, every mythology, every tribal narrative contains tales of both extreme longevity and perpetual youth—of those who live, like gods, forever, and those who stay forever young. In the *Epic of Gilgamesh*, for example, the world's oldest surviving literary work, a great Sumerian king searches for the secret of immortality before coming home, chastened, to die.[20] In ancient Greek texts, a people known as the Macrobians were said to wash in a spring that smells of violets and live to 120. And in the sixteenth century, Spanish explorers led by Juan Ponce de León searched "every river, brook, lagoon or pool" across Florida and the Bahamas in pursuit of a magical fountain rumored to slow, or at least conceal, the passage of time.[21]

Their quests, of course, came to naught. But today, armed with actual, workable tools, scientists and technologists are beginning at last to crack the underlying codes of both life and death. They are starting to understand, at a molecular level, how our bodies die and what might slow that process. They are building microscopic machines that can be swapped into a human body as its own systems start to fail. And, most audaciously perhaps, they are learning how to unravel the strands of human consciousness, exploring the brain and the mind as digital networks capable of being stored, uploaded, and rebooted.

To understand the approaches that these problem solvers are taking, it is helpful to dwell—even briefly—on the nature of death. When we die, like any other living organism, our physical being ceases to function. Our breathing stops and our hearts shut down. Our eyes and ears and lips and legs no longer do their work. Accordingly, one way to beat death is to prolong the physical working of these bodily parts—not only to engineer longer-living lungs and hearts and eyes, but to hack the entire body at a cellular level, manipulating the biological processes that cause its parts to age and decay.

Another way, though, is to see death as the loss of thinking, of the memories and jokes and convictions and ideas that define us as individuals. This is the part of our bodily selves that doesn't necessarily rest in or upon our bodies, the ever-present shadow that religion and philosophy generally label the soul. It is what we miss the most when a loved one dies. And it is the part that is theoretically amenable to digital re-creation.

These two approaches are by no means contradictory. In fact, both generally share a deep and abiding reverence for Moore's law— Gordon Moore's famous 1965 dictum that computing power doubles every two years—and a belief that virtually everything that humans now do through biological processes can eventually be done more quickly and efficiently by computers.[22] But the technology they would each require and the vision of immortality they imagine are fundamentally different. One group—I'll call them the body hackers—is tackling the problem of death at a biological, physical level, trying to prolong and extend the living body for as long as possible. The other—I'll call them the animators—is abandoning the body in favor of the soul, searching for digital ways to preserve or re-create an individual's brain long after their physical form has ceased to exist.

The Body Hackers

In many ways, the medical thinkers and researchers who populate this field are the logical successors to Ponce de León, the seekers, still, of a magical elixir to turn the tide against death. Rather than searching for an external salve, though, this group starts where life ends: in the intricate biology of the body itself. In particular, the body hackers have focused on a range of biochemical switches, any one of which, in theory, might be able to stem the process of aging.

The most obvious target are telomeres, since they are so deeply implicated in the problem of senescence. Most cells in the human body divide fifty to seventy times over the course of their life; at each division, the end of the cell—the telomere—shrinks by another notch, until it becomes so small that no further division is possible. At that point, the cell enters senescence and the body part to which it is connected begins its inevitable decline. But if a body's telomeres could somehow be preserved, the logic runs, then that body's cells could theoretically reproduce as vigorously in old age as they do in youth; the cells, in other words, could become immortal.[23] "My simple view," writes one researcher, "is that aging is those things that go wrong when cells lose their ability to divide . . . If we could replace our cells as rapidly as they deteriorate, we could probably live very long, if not indefinitely."[24]

Such a happy state of affairs already exists in a handful of other living creatures. The so-called immortal jellyfish, for example, can transform itself back into a polyp, its earliest stage of life, at any subsequent point in its development.[25] Some deep-sea sponges can live for hundreds or thousands of years, constantly producing new generations of healthy stem cells.[26] *C. elegans*, by contrast, a tiny and well-studied worm, is normally quite mortal, living on average for only twenty days. But by tinkering with one of the genes that seem to regulate the animals' life span, scientists have bred *C. elegans* that live for twice as long; by removing the worms' reproductive systems, they have produced individuals that live six times longer than normal—the equivalent in human terms of five hundred years.[27] In mice, an enzyme called telomerase has been shown to prevent shortening of the telomeres and thus extend cell life span; when researchers at Harvard Medical School treated genetically modified mice by reactivating their ability to produce telomerase, the once-feeble and elderly mice were visibly rejuvenated. Their spleens and testes grew larger again, and their brains sprouted new neurons. "What we saw in these animals was not a slowing down or stabilization of the aging process," reported the study's lead researcher. "We saw a dramatic reversal."[28]

Extending these benefits to humans is not, in theory again, particularly difficult to contemplate. If telomerase could be introduced appropriately into the body, it could be used as a preventive treatment for aging, dousing an individual's cells before they begin the process of senescence.[29] Individuals in search of an extended life, or even just a healthier old age, could dose themselves with telomerase-related compounds, or with chemicals engineered to prevent their telomeres from shortening; already, some adherents are trying to mimic telomerase's effects with a supplement derived from dried astragalus root, a plant often used in traditional Chinese medicine.[30] To date, though, their efforts have been foiled by two underlying problems. First, as one prominent aging researcher has noted, "mice are not little men"—their cells produce telomerase throughout their lives, while those of humans do not.[31] And second, raising the level of telomerase in humans could unleash a Faustian biological bargain, since infinitely replicating cells are by their very nature cancerous.

Meanwhile, teams of researchers around the world are probing and experimenting with a growing cast of other proteins and enzymes, slowly unraveling the cascade of biochemistry that both triggers and accompanies aging. There is a class of proteins called sirtuins, for instance, that showed promise in the early 2000s of being able to turn off entire sections of an organism's genome and slow its metabolic rate.[32] Metformin, a drug used primarily to lower blood sugar levels in patients with type 2 diabetes, has also shown tantalizing signs of slowing the progression of both heart disease and cancer.[33] Intriguingly, both sirtuins and metformin—along with a range of other substances—operate within the body in ways that mimic starvation. Simply put, if an organism is desperate for nutrients, its cells seem to retreat into a biochemically defensive state, shuttering every function—including aging—that is not needed for its immediate survival. In another bit of biological irony, therefore, nearly dying seems to forestall aging, at least in those species where experiments have thus far been conducted. Flies subjected to extreme dietary restriction generally live twice as long as better-fed creatures and demonstrate an extended ability to fly. Mice live 30 to 50 percent longer and have lower rates of cancer, diabetes, and other degenerative diseases.[34] If humans respond similarly—slowing all metabolic systems to respond to what feels like starvation—we could theoretically prolong life for another few hungry years.

Thus far, none of these approaches has come close to identifying a single switch or genetic pathway that might easily be plumped into action. But the prospect of prolonging life has already been far too heady—personally and financially—for investors to ignore. In addition to the slew of pharmaceutical firms tinkering with sirtuins and telomeres and metformin and more, individual investors like Google's Sergey Brin and PayPal's Peter Thiel are pouring money into life-extending experiments, often with the backing of purpose-built ventures like Google's Calico (short for the California Life Company) and Human Longevity Incorporated (cofounded by Craig Venter, whose earlier company sequenced the human genome in 1998). Armed with the conviction of men (usually) who made their fortunes from mixing math with technology, the body hackers believe that life, too, can be deconstructed and rearranged, broken into its component pieces and resurrected into something better. Something fitter. Something engineered to run forever. "I have the idea that aging is plastic, that it's encoded," says Joon Yun, a doctor who runs a health care hedge fund. "If something is encoded, you can crack the code. If you can crack the code, you can hack the code!"[35] His views and optimism are echoed by Aubrey de Grey, a preeminent and controversial aging researcher who insists that aging "is not something inherently mysterious, beyond our power to fathom." Instead, "aging of the body, just like aging of a car or house, is merely a maintenance problem."[36] And if it's merely a maintenance problem, he continues, "You could stop thinking about aging as a hopelessly complex *theoretical* problem to solve, and get on with attacking it head-on, as an *engineering challenge* that needed to be overcome."[37]

In 2001, de Grey cofounded the Methuselah Foundation to invest in and promote life-extending technologies. The foundation's public mission statement makes its goals clear: "to make 90 the new 50 by 2030."[38]

The Animators

The second path toward ending death is even more extreme. It springs from the ancient conviction that humans are more than their bodily selves—indeed, that the part of us that exists beyond our hearts and cells and telomeres may be what makes us entirely human and uniquely us. If we are, as the seventeenth-century philosopher René Descartes famously wrote, "distinct from our bodies," then we needn't

Domenico Zampieri, *The Assumption of Mary Magdalene into Heaven*, 1620

go through the messy exercise of preserving flesh and bones to attain immortality.[39] We just need to save our minds. Why bother to hack the body, this logic runs, if you can instead store and reanimate the self?

Like the body hackers' origins in myths of eternal youth, the quest for reanimation runs long and deep—so deep, in fact, that one can see its misty origins in nearly all the world's religions. We know, and have always known, that humans die. But we have also hoped, and perhaps have always hoped, that some invisible piece of us lives on—in the clouds, in the seas, on the altars of our grandchildren, or beyond the pearly gates. As early as 360 B.C., Plato proposed that, because life and death were opposites, and generated from one another, the souls of the dying were destined to be reborn. "Revival," he surmised, "is the birth of the dead into the world of the living."[40] Some fifteen hundred years later, Thomas Aquinas refined this concept and made it integral to Christianity, arguing for the existence of a human soul separate from the body, something "called intellect or mind . . . something incorporeal and subsistent."[41] Descartes then formalized these ideas into a clear-cut duality: what we think of as "I," he hypothesized, is really two. There is the bodily me, which

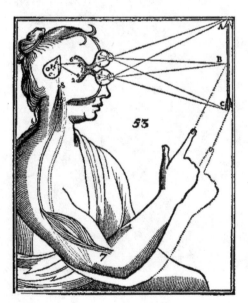

Descartes's illustration of dualism. The body's sensory organs pass inputs into the brain, and from there to the immaterial spirit.

feels and senses and moves. And then there is the actual me, the thinking me, which directs the body and interacts with it, but remains both separate and divisible.[42]

For over two thousand years, however, belief in this un-bodily ascent has been just that: a belief. A fervently held one in many cases, a much-flourished and celebrated and beautifully described one. But a belief nevertheless, without a glimmer of scientific proof or mechanical explanation.

Outside a small village in northern Vermont, however, that disconnect may be poised to change.

Martine Rothblatt is a relative newcomer to the world of digital reanimation—or revival, as she prefers to call it. Born a man and trained as a lawyer, Martine spent the early parts of her career working in the then-emerging field of satellite communications. She helped launch Geostar, a pioneering car-navigation system in the early 1980s; founded what would become SiriusXM; and, at a party one day, fell madly in love with a woman named Bina. She and Bina married, and settled down to build their lives with their four children.[43]

In 1994, Martine (still then Martin) decided to undergo gender confirmation surgery, transitioning into a female body and identity. She and Bina stayed together, happily embracing an evolving sexuality for them both; Bina calls herself neither straight nor gay, but "Martine-sexual."[44] When their seven-year-old daughter, Jenesis, was diagnosed with a rare and life-threatening disease called primary pulmonary hypertension, Martine used some of her proceeds from Sirius to launch a new firm, United Therapeutics Corporation (UTC), to invest in the drugs that might save her. By 1999, UTC had gone public and won FDA approval for a new and powerful drug. In the process, and because she had earlier linked her salary to the company's stock price, Martine became the highest-paid female CEO in the United States.[45]

For most people, this—a happy marriage, four healthy children, two wildly successful companies, and a sex change—would have been enough. But Martine Rothblatt is not most people. Instead, awed by the power of the love she felt for Bina, and worried about the inevitable prospect of losing her, Martine decided that they should both live for-ever. Or at least that some form of their love and their lives should be able to extend beyond their worldly selves. "If we could create digital minds," she started to imagine, "we could continue this love affair forever."[46]

That's what's brought me, on an achingly beautiful fall day, to the tiny village of Lincoln, Vermont—population 1,271, plus an unidenti-fied number of sheep and one robotic head. The roads are quiet and covered with fallen leaves; the town has a general store and a library, sprawled along a rocky stream.

It's the kind of place where you wouldn't mind spending forever.

Which is exactly what Martine has set out to do. In a nondescript garage tucked away in the hills, she oversees the Terasem Foundation, a nonprofit devoted to extending lives through digital preservation. On the foundation's related website, Lifenaut.com, users can upload their "mindfiles," a digital database of videos, images, social media posts, and other personal documents that can subsequently be saved, searched, and downloaded. In the garage outside Lincoln, Bruce Duncan, the foundation's director, is tinkering with a robotic bust of Bina, uploading her mindfiles into an early version of what Martine hopes will eventually become a "mind clone."[47]

When I visited in October 2018, the robot was still distinctly clunky. Her facial movements, while impressive, looked forced. Her answers were flat. When I asked her, at Bruce's prompting, to tell me about just war theory (she had recently returned from a trip to the U.S. Military Academy), she gave a fine, if somewhat canned, textbook response. When I asked about her children, she blankly repeated the basics: four kids, all loved. But then, eerily, when I asked about her father, she be-came agitated, telling me considerably more than I had asked for or needed to know.

Sitting there, it was hard not to imagine I was seeing the future. No, Bina doesn't feel human yet, in any serious way. But she does feel like a presence of sorts, a Model T version of what will someday be more

Bina48, October 2018

polished, and more real. Her body (or torso, I guess, more precisely) wasn't what intrigued me, since it's still too awkward and uncanny. It was, in Cartesian terms, her mind—some glimmer of intelligence and personality breaking out of the code that was animating her. Some soul that somehow knew she didn't like her father.

In her customary all-encompassing fashion, Martine has built a mini-empire around her visions of Bina. There is the robot itself, of course, and the website. There is the Terasem Foundation and a religion (also known as Terasem) with its own ashrams.[48] At its core, though, Terasem's vision of what it and others refer to as a "transhumanist" future is based on science and technology, and on the evolving field of artificial intelligence. Specifically, it is based on rapidly emerging work that suggests that (a) human brains function largely like computers; (b) human thought patterns can thus be replicated by advanced computing software; and (c) such software-based thought can eventually approximate and replicate human consciousness.[49] We will merge, in other words, with our software-based minds.

One of the most vocal advocates for this view is Ray Kurzweil, a prolific author, futurist, and director of engineering at Google.[50] In 1999,

Kurzweil wrote a book titled *The Age of Spiritual Machines*, famously proclaiming that machines were well on their way to becoming more intelligent than humans.[51] By the year 2020, he predicted, a massively powerful neural net computer would be able to perform about 20 million billion connections per second—roughly equivalent to the thinking capacity of a human brain. By 2023, a silicon brain equivalent will run more than a billion times faster than its human predecessor, for a cost of only about a thousand dollars. By 2060, and following the inexorable and exponential march of Moore's law, a single personal computer will replicate the brainpower of a trillion human brains.[52] In the process, Kurzweil argues, these massively powerful thinking machines will develop what we see as free will. They will experience something akin to spirituality.[53] Most important, they will evolve alongside their human progenitors, eventually merging and becoming one with them. In less than one hundred years, he asserts, "there won't be a distinction between human and technology. *This is not because humans will have become what we think of as machines today, but rather machines will have progressed to be like humans and beyond.*"[54]

In the years since *Spiritual Machines* was first published, plenty of critics have disputed both Kurzweil's timeline and his underlying hypothesis. Philosophers have critiqued his definition of consciousness; technologists have questioned his unflagging belief in the law of accelerating returns.[55] But researchers have also been steadily plugging away at most of the developments he foresaw and, in particular, at the kind of human-mind transcendence that Martine is trying to make work.

Central to these endeavors is the quest to map the human brain; to understand, at a molecular level, how its billions of cells and trillions of synapses connect and work together.[56] It is a herculean task. But, much as Kurzweil predicted, the exponential growth of computing power has indeed made it possible for researchers to know more and more about the brain, at faster and faster speeds. Using silicon-based chips, scientists at the University of Southern California have built integrated circuits that precisely mimic the processing patterns of clusters of human neurons.[57]

At Harvard's Center for Brain Sciences, neuroscientists have described how a chemically fixed and plastic-embedded brain could be sliced into microscopic pillars of tissues, each of which could be digitally imaged using a highly specialized ion beam electron microscope. Given sufficient computing power, the pillars could then be stitched together, creating a complete digital map of all the neural connections in the original brain—what researchers in this field refer to as the "connectome." "Sectioning an entire brain into 20-micron-thick strips with 100% reliability will be a difficult engineering challenge," the head of the team modestly admits. "But one that appears quite achievable."[58]

Perhaps. Like Kurzweil, many researchers in this field take inspiration from the Human Genome Project, which was launched to great fanfare in 1990 and completed ahead of schedule in 2003.[59] In that effort, researchers used supercomputers to painstakingly dissect the more than three billion nucleotides contained in a representative set of human chromosomes. When the project began, researchers presumed it would take at least fifteen to twenty years to complete. As computing power accelerated over the period, however, so did the pace of discovery: the first draft of the genome was finished by 2000, and the final version published in 2003. Even more impressively, the time and money required to identify a specific gene had plummeted. When James Watson had his own DNA sequenced in 2007, it cost nearly $1 million and took two months; by 2017, commercial firms were promising to sequence any individual's genome for less than $100 and in under an hour.[60]

If processing speeds were to accelerate at similar rates over the next few decades (as most computer scientists predict they will), and software were to improve to a point where algorithms were essentially reprogramming themselves (again, a widespread prediction),[61] then a potential Human Brain Mapping Project could theoretically have sufficient data-processing power to dissect and depict the billions of neurons and trillions of connections that together make a mind.[62] At that point, there are two paths that scientists could pursue. They could build a brain from scratch, or, more likely (and less scarily), they could create the scaffolding upon which to re-create or reanimate one that already exists. To quote Kurzweil again: "Perhaps more interesting than this scanning-the-brain-to-understand-it approach would be scanning the brain for the purpose of downloading it. We would map the locations, interconnections and contents of all the neurons, synapses and neurotransmitter concentrations. The entire organization, including the brain's memory, would then be re-created on a digital-analog computer."[63]

Which is where Bina or her descendants reenter the picture. Imagine that someday all the tangible artifacts that Bina Rothblatt created over the course of her life—her photos, videos, blog posts, and so forth—could be merged with a software system that replicated the inner workings of her brain.[64] Imagine then that after Bina died, her existing memories could be used to generate new ones, new ways of talking and interacting and learning. Imagine her growing as a person over time—well, not a person exactly, but an entity, maybe an avatar—that felt to the person on the other end of the relationship more or less like the original Bina. You could put a body on her, too—Kurzweil is planning to build new bodies, eventually, for both himself and his father—but for many people, including Martine, the brain is more than enough.[65] "The most practical way to transcend death," she patiently explained to me, "is to take the essence of you, and take it out of the body, and put it somewhere that it won't fall apart."[66]

There is a reason, perhaps, why someone decided to call our collective digital storage system "the cloud."

To be sure, the future that Kurzweil and Rothblatt are so confidently contemplating is a long way off. It relies on an acceleration of computing power that may never quite happen, and on an organization of brain matter that could prove far trickier to unlock. We could stumble into nuclear war before this line of research comes to fruition, or be distracted by more immediate concerns. But in many ways, the search for digital immortality is proceeding exactly as Kurzweil first predicted: humans are building smarter and smarter machines, at faster and faster rates, aiming toward a point at which our machines get smarter than we are, and we become one with them. Kurzweil calls this the singularity.[67] Martine refers to our transhuman future, a time at which the "software-based mind" will become "a techno-immortalized continuation of the person's identity."[68] It's not quite immortality as the ancients would have described it. But it's starting to get close.

When *Les Temps Perdu* Were Cast in Stone

Your love grows old in parts. One morning, the person you've lived with and adored for years is the same as they've ever been. The next, they have aged in bits. An elbow that sags a small window of skin. A neck that grows rougher. The body betrays in painful ways, stealing dignity along with youth. "What of soul was left, I wonder," wrote the poet Robert Browning, "when the kissing had to stop?"[69] No wonder that we have always grasped for some magical escape. Any way to slow time's toll and wrestle back the bodies that were lost.

Until recently, though, such quests were either fantastical or theoretical. Now they're becoming real, and posing questions that threaten to leap quickly outside the realm of abstract philosophy.

What will we do—as a species, as a society, and as individuals—as technology offers the promise of longer and longer lives? How will the possibility of a transhuman future shape how we structure our more mortal lives today? And what happens to humanity if and as we figure ways to regain what was once lost?

These are huge questions, so I will start with the part that seems easiest: the bodies. History would suggest that none of the life-enhancing techniques that are currently under investigation will prove, in the end, to "cure" mortality in any meaningful way. They are in the early stages, still, and dedicated, as early-stage technologies must be, to probing connections rather than providing cures. But the underlying science is getting better and better, closer and closer to unlocking at least some of the core processes that drive and define aging. Metformin, telomerase, another potentially powerful drug called rapamycin—all show real, albeit early, evidence of the ability to reverse or slow the biological pace of aging. As scientists unpack their biochemical pathways, people will be increasingly tempted to treat aging like mosquitoes or the common cold—as something to be warded off as much as possible. Already, many of the scientists involved in this work dose themselves heavily with metformin or antioxidants or various growth hormones. Ray Kurzweil apparently takes ninety supplements a day.[70] Peter Thiel reportedly has a pill butler to serve and keep track of the drugs he takes. As soon as there are any proven means to delay aging or its attributes, people will gobble them up. We always do.

And the rich will gobble faster and more. Indeed, it seems obvious that the development of age-defying technologies—be they drugs, modified organs, or genetically reengineered cells—will further exacerbate the wealth-related health asymmetries that exist today. People in wealthier countries already have considerably longer life spans than those in poor ones (eighty-one years in Denmark, for example, versus fifty-three in Chad). Wealthy people in any given country live longer than their poorer compatriots.[71] As new medical technologies emerge— none of which are technically lifesaving under our current definition of that word—their usage will almost certainly be clustered around those most able to afford them.[72]

These growing ranks of the wealthy old are likely to wield a disproportionate amount of power, lingering longer in political office and voting against any redistribution of wealth toward younger generations. Surveying the early vistas of this future from Japan, where life expectancy has already hit eighty-four and more than 20 percent of the population is older than seventy, Gregg Easterbrook notes that young people there have some of the world's worst voter-participation rates. He quotes a specialist on Japan as saying, "They think the old have the system so rigged in their favor, there's no point in political activity."[73] Such trends are almost certain to continue and expand. So, too, are demands for an extensive array of caregivers to tend to the needs of those who stay old for a really long time. Private chefs and housekeepers, for example.

Physical therapists and nutritionists and home health attendants. Even if some of these providers are robots (and they will be), one can easily envision an ever more stratified world, torn between those with the means to grow wealthy and old, and those consigned to take care of them.

Meanwhile, and somewhat paradoxically, living longer might shorten some of our most intimate relationships. Marriage, after all, is a vestige of the Neolithic era, born at a time when life spans rarely exceeded forty years. It is, as I've argued repeatedly, a social structure tied to a distinct biological rhythm. We are born; we become fertile; we mate; we die. But what if that dying occurs decades after the mating and rearing are done? What if we can mate over a wider range of time and through a variety of nonsexual means? Will we still commit as passionately to the "till death do us part" contract?

Probably not.

Instead—and alongside the parallel developments in online communication, reproductive technologies, and robotics described earlier in this book—we will likely move to more episodic arrangements—to shorter commitments, fluid partnerships, and more elastic notions of love. Some of us will no doubt still fall in love in ways that last forever. But there have always been out-of-wedlock romances in a world defined by monogamy and happily independent singles amid a sea of married folks. What's changing, I'm arguing, are the defining structures of our social lives, the norms that quietly regulate what we'd like to think of as choice.

In some odd way, it feels that we have always known, or at least dreamed about, this shift toward immortality. Every one of the world's major religions, after all, has its precipice of transcendence, its moment when man not only reveres the gods, but becomes one of them. Every poet, every song, every one of us yearns to bring back what was lost, and to retain some element of ourselves and our loves forever. "Don't go" is perhaps the most common refrain across our verbal culture. "Please stay."

And now technology has created at least the prospect of making this transcendence real. If Kurzweil and his colleagues are right, if Martine has created in Bina48 an early prototype of what techno-immortality might entail, then we are indeed on the cusp of a monumental revolution, bigger, perhaps, than any our species has ever known. Extending youth along the lines of what the body hackers describe will shift our societies and social patterns; re-creating our minds through digital means will fundamentally alter our species.

But let's pause for a moment on "re-creating our minds." Bodies, again, are easy. We can contemplate what it means to make an artificial eye, or a genetically engineered cell. Brains are much harder to bend our minds around. And that's where many discussions of artificial intelligence and digital immortality get stuck. Can even the smartest of smart machines really *think?* the philosophers wonder. Would a thoroughly convincing version of Bina48 really be an immortal offspring of the biological Bina, or just a clever replica? Even Martine has posed this question for Terasem, arguing that software-based minds should not be considered conscious until a panel of outside experts, reviewing both the brain-based mind and its software-based offspring, are able "preponderantly" to conclude that "the purported consciousness did in fact appear to be equivalent to the consciousness of the subjects who created the mindfiles."[74]

This may be so. But I would argue that the philosophical question of consciousness is in many ways less important than the practical one of impact. Bina48 is real if Martine and those who love her think that she's real. She's real—any robot is real—if we bring it into our lives and re-shape our behavior accordingly. If something like our brains can be re-created computationally, if something like our selves can exist forever online, then we will embrace these machine-man things and bring them into our lives, regardless of whether or not we technically see them as sentient. To give a trivial example: many of us have brought Siri into our cars already, and Alexa into our homes. Are they conscious? No. But they're real. Now take it up an emotional notch. Your best friend dies, or a beloved parent. You know they're gone, mortal. But what if you could speak to them, as you typically did in real life, on your phone? What if they still texted you, and sent photographs to share? What if you could call them at the end of a rough day and tell them what went wrong?

You'd do it, right? So would I. And philosophical questions of consciousness would likely fall prey to the more immediate and all-too-human longing to connect.

This, I think, is where the first wave of techno-immortality will take us: to a place where some semblance of our lives and loves can be recaptured in ways that feel pretty close to real. Already, every day, we are digitizing our lives: pouring our thoughts out on Twitter, posting our photos on Instagram, leaving footprints of every trip, meal, and hookup littered across our smartphones. Should they ever be interested, my grandchildren could learn far more about me than I ever could about my grandparents—far more, indeed, than they could ever need or want to know. My children can find their notes to me and mine to them; my husband can retrieve decades of intimate texts, wrapped up like letters used to be, and waiting in the cloud. Like nearly everyone living in the opening decades of the twenty-first century, I am carving a digital legacy of my time on this planet, one that will endure far longer and more reliably than memory alone could ever have provided.

In the end, it doesn't matter so much who interacts with this material and whether my story is animated by a sexy avatar, a convincing robot, or a string of zeros and ones. What matters is that I'm telling my story, and capturing my story, in ways that feel accessible to others and specific to me. Because that, after all, is what humans have been striving for ever since we began to draw pictures on cave walls and letters in the sand. Not necessarily to live forever in a bodily form, but to create stories—legends, sagas, legacies—that live beyond us. To leave something behind that remembers our presence and our moment on earth. *L'shana tovah*, reads the Jewish blessing on the Day of Atonement, "may her name be inscribed in the Book of Life." May her memory—his memory, their deeds, our union—live on and on forever. We fear death because we fear disappearing, because, as Freud wrote, "in the unconscious every one of us is convinced of his own immortality."[75] Throughout human history, most people's memories were kept alive for a generation or two, just long enough for their immediate descendants to remember their names and their lives. Thus the importance and poignancy of rituals like Mexico's Day of the Dead, urging the living to keep the dead alive by recalling their names and honoring their earthly existence.

But across this history, memories could only ever last so long, leaving generations with the existential puzzle of *why*. Why bother being on this earth if your journey is quick and unremembered? Now, whisper the body hackers, we can change that. By delving deep into our own biochemistry, we can fix the flaws that evolution left, and extend the span of life to something larger, something longer and more fitting to the prowess of our species. And perhaps they shall. By approaching the puzzle of the mind, however, the animators are actually tackling a greater quest—one even more arrogant and profound. Because if we are our minds and our souls—the Cartesian beings-because-we-think—then the bodily part of our lives isn't that important at all. Instead, if we want to beat mortality, we need only ensure what our ancestors long yearned for: that our stories and our memories endure.

Today, technology is on the verge of making that leap. It may happen through mind clones or something simpler, through digitally downloaded brains or websites that store our thoughts in their clouds. We may learn in time how to upload our digitized memories to newly configured bodies, or instead content ourselves with inanimate replicas that echo at least some fragment of the person we loved—a toddler laughing and singing on something like FaceTime, a lover sending texts to your phone. The precise tools in the end may not matter. Because if what makes us ultimately human is the knowledge of our deaths, then immortality means knowing—dreaming, believing, trusting—that we will live on. And once we have that belief, by definition, we stop being as human as we once thought we were.

Conclusions
Welcome to Tomorrowland

Will robots inherit the earth? Yes, but they will be our children.
—Marvin Minsky, 1994

All predictions are wrong, that's one of the few certainties granted to mankind.
—Milan Kundera, *Ignorance*, 2000

Shulamith Firestone died poignantly. Once heralded as the new voice of feminism, the Canadian-born scholar was sixty-seven when her landlord found her, alone in her fifth-floor tenement in New York's East Village. There was no food in the apartment, and it appeared she had starved to death.

Several decades earlier, before she had even turned twenty-five, the younger Firestone had been a radical, a leading voice for revolution during a distinctly revolutionary time. She was part of a broader struggle—the struggle that twisted and forced the civil rights and anti-war movements of the 1960s and '70s to consider women's rights as well—but she was also, and always, alone, stretching further in her theories and proposals than most of her comrades would dare. From the beginning, her ideas were intense, pocked in equal measure with analysis and fervor. In retrospect, she may long have been skirting, like so many geniuses, the edge of madness. And it was madness, in the end, that brought her down.

In 1970, though, before schizophrenia took its toll and dragged her into the solitude that would ultimately kill her, Firestone wrote a brilliant book called *The Dialectic of Sex*. It was a complicated read—dense, theoretical, angry, and far too radical to be taken seriously in most mainstream circles. But read fifty years later, in a time when both technology and social norms have evolved further than even Firestone dared to imagine, *The Dialectic of Sex* seems oddly, eerily prescient.

Arguing from a rigidly Marxist perspective, Firestone declared in her *Dialectic* that the first inequity of human civilization, and the worst, was the division of the sexes. The family, she wrote, channeling Engels but with a distinctly sharper edge, is the "vinculum through which the psychology of power can always be smuggled." Women are enslaved to men, and children to adults, simply because reproductive biology demands it, giving paternal authorities—that is, wealthy men—the power to oppress everyone else. Like most hard-core Marxists, Firestone was better on the diagnostic side than the prescriptive, sharper at highlighting the ills brought upon humanity than suggesting any means, short of revolution, of redressing them. But in sketching a way forward, she stumbled upon and connected two unlikely trends: assisted reproduction and information technology, or what she labeled, in the early days of the 1970s, "cybernetics."

These two developments, she predicted, would break the social structures that had persisted since the dawn of civilization. They would bring the revolution. Because once humankind could replace itself through artificial means, she asserted, women would be freed from the biological constraints that had chained them for so long to their children and families. And once machines replaced the work that men historically did, they, too, would be freed from having to build their lives around labor. Together, new means of production and reproduction would blast through the structures of marriage and family, Firestone proclaimed, paving the way toward freedom. Or as she put it, "Now, for the first time in history, technology has created real preconditions for overthrowing these oppressive 'natural' conditions . . . The double curse that man should till the soil by the sweat of his brow and that woman should bear in pain and travail [will] be lifted through technology to make humane living for the first time a possibility."[1]

Clearly, this world has not yet come to pass. It probably never will. Indeed, fifty years after the *Dialectic*, fifty years after the feminist, civil rights, and antiwar movements, humanity is as entangled in conflict as it ever was—fighting, killing, over race and religion and land. And yet. At the very core of her argument, Firestone was right, and we are living the proof. Because the history of humankind is indeed marked by technological change and social response, by machines and humankind progressing together, moving not quite in tandem, but in a hobbled-leg race in which one piece leans forward and the other, stumbling sometimes but without fail, follows suit. Man makes the plow and then the plow remakes man, pushing him to create property, and marriage, and tiny social formations known as families. Man makes an engine that runs on steam, and before long all men are rushing at a faster pace, connecting their lives to an industrial core whose fragments touch everything: how children are raised, how men earn their wages, how women are set to behave. Man invents the car, and the pill, and the dishwasher, and watches as they change, in turn, how he works and loves. Woman hasn't invented much of anything over these millennia, but this, too, is driven by technology, and the extent to which—at least until recently—it has remained the province of he-who-labors-outside-the-home.

Today, we are living through another of history's moments of radical change. If the Neolithic Revolution thrust humanity into what we now think of as civilization, and the Industrial Revolution into a world dominated by machines and commerce and speed, then the Digital Revolution—or whatever our descendants will eventually label the transformational technologies wrought by computing and digitization—will yank us into a postindustrial world, with patterns of life, work, and family formation that are strikingly different from those that prevailed in the past.

Already, the inklings of that future are seeping into everyday life. We organize our days and dreams on devices we call smartphones, regularly confiding our secrets to these computers that cling to our hips and seeking their counsel about where to go, what to do, and how to get there. We meet our mates in cyberspace and sometimes have sex with them there, too. We are using hormones to transfigure our bodies and genetic engineering to craft our children. Yet, frequently, strangely, we compartmentalize these shifts into things-that-are-technical and things-that-are-personal, failing to see the intricate and intimate links between them. Because when we start treating our computers as confidants, we change a bit in the process, too. When smartphone apps can substitute for matchmakers and computers simulate sex, they leap from the realm of gadgetry to become something more profound, something that shapes not only how we interact with our devices, but how we relate to other humans as well. When we change our tools in fundamental ways, in other words, the changes radiate far beyond the wholly technical realm, beaming into our bedrooms, our boardrooms, our nurseries, and beyond.

Specifically, the digital revolution of the twenty-first century is poised now to reshape four fundamental and linked areas of human behavior: how we work, how we mate, how we marry, and how we love.

The Future of Work

Let's begin with work, the most obvious of these changes. As numerous accounts have already predicted, the digital revolution is poised to upend structures of labor that have prevailed since its industrial predecessor. Over the next several decades, as robotics, artificial intelligence, machine learning, and nanobots storm out of the laboratory and into the factories and offices of the future, the workplace as we currently know it will be transformed. Assembly-line workers will be replaced by smart machines and robotic appendages. Drivers—of taxis, trucks, trains, and planes—will cede their place to drones and driverless vehicles. Back-office functions such as purchasing, payroll, and accounting will effectively move from the cubicle to the cloud, taking what was once a mostly human labor force along with them. Increasingly, even high-touch fields like teaching and medicine will be augmented and at least partly supplanted by digital intelligence—students will study and learn online, as they already are doing; patients will be scanned, examined, and diagnosed by intelligent robots uploaded with complex algorithms.[2]

The nature of work is always changing, and always has been, from the gatherers who became farmers in the Neolithic Age to the sea captains who morphed more recently into astronauts. As technology reshapes the dominant means of production, the producers are inevitably dragged along: those with the power and resources (James Watt, Henry Ford, Elon Musk) seize and re-create the means themselves; those who labor for wages adapt to the jobs that pay. Such was the case as steam-powered engines thrust their way across England's once-agricultural economy: farming men became factory workers who labored at the loom. So, too, with the development of cars, which drew generations of workers into patterns of assembly-line production that quickly felt eternal. And such will be the case again with digital technologies—the labor force will need to adapt, throwing millions and millions of existing workers into inevitable obsolescence. From the perspective of an individual worker, the change feels wrenching, even tragic. Across the broader sweep of history, though, that's just what technology does.

This time, however, the transformation promises to be particularly profound. Because unlike anything that came before it, this round of technological change has the potential to make work as we know it virtually extinct; this time, the jobs that are being destroyed by progressive waves of automation show little sign of being re-created elsewhere. Instead, the sheer efficiency of computer-aided manufacture means that machines, on their own and without a lot of human intervention, will soon be able to produce most of humanity's basic needs. Food—a single farm in Iowa, manned entirely by robotic milking machines, can produce fifteen hundred gallons of milk a day.[3] Clothing—a new sewing robot (or "sewbot") can produce over eleven hundred T-shirts in an eight-hour period, or roughly as many as seventeen humans.[4] And whereas General Motors employed more than 618,000 workers in 1979, Tesla today has only 46,000. Admittedly, these ratios differ significantly around the globe: in still-rural countries such as Chad, 75 percent of the country's working-age population remains tied to the land and to food production, whereas in postindustrial Estonia, it's only 3 percent. But as information technologies and commercial forces push inexorably around the globe, the trajectory of digital technologies—of smart machines built to replace and improve upon human labor—will expand across all but the most isolated areas. Within a few decades, we will reach the tipping point that Keynes predicted in 1930: technology has made our society so economically efficient that there are simply not enough jobs to go around.

Meanwhile, even the jobs that remain in this new world are increasingly being configured in fundamentally different ways.[5] They are less stable than the jobs of the past, less male-dominated, and generally untethered from the once-standard nine-to-five day. As of 2018, more than 50 million people in the United States and 110 million in China were employed in the so-called gig economy, working essentially as day laborers for services such as Uber, TaskRabbit, and Didi Chuxing.[6] Millions more were selling products that ranged from hand-knit booties to makeup tutorials on sites like Etsy, eBay, and Alibaba. None of these workers are employed under what were once commonplace norms: they don't have contracts per se, or regular working hours. They don't have performance reviews, health insurance, or unions. Unlike the wage laborers of the industrial era, these information-age workers don't have any kind of long-term relationship with the entity that structures their livelihood. They work, instead, on their own and in a piecemeal fashion, stitching together what once were salaries from a series of tasks and gigs.

Not all contemporary workers inhabit the gig economy, and not all employers seek to emulate its ways. But as gig work becomes more prevalent, its patterns are almost certain to seep into more traditional workplaces as well, nudging both employees and employers to dismantle key components of their older labor contract. In the heyday of the manufacturing economy, a worker for General Electric or Siemens would typically sign on in his youth, expecting to remain at the same firm (and often the same plant) for the full stretch of a roughly forty-year career. He (and it was usually a he) would work for a fixed number of hours at a well-described job, earning wages and vacation days and retirement benefits that were set in advance and negotiated to apply across an entire tier of workers. Now, pushed by the ways and means of gig work, all these arrangements are beginning to fray. In place of defined-benefit pension plans, workers are signing up for various self-funded and less remunerative retirement schemes;[7] instead of negotiated overtime, they are agreeing to just-in-time schedules that seek to match an employer's precise labor supply with its projected demand.[8] Across the postindustrial world, labor is becoming more personal and idiosyncratic; individuals build their brands (if they're lucky) and their career paths, rather than, as was the case in the past, joining the ranks of something—a firm, a union, a political party, or a military unit—bigger and more permanent.

Once again, it is tempting to attribute these shifts to cultural evolution, or to the adult ways of a millennial generation raised to be mobile and self-centered.[9] But millennials aren't pursuing gig-economy jobs because they prize flexibility and some illusion of work-life balance; they're not pivoting from post to post purely in search of more generous snack bars and whiffs of an IPO. No, they're behaving this way because technology made them. More specifically, they are responding logically to the technological environment in which they were raised, and to the technologically determined economy that confronts and confines them. This is the generation, after all, that came of age with smartphones, a generation accustomed to Netflix and YouTube and a constant infinity of choice. Many of them are naturally going to favor the same range of ever-present options in their work lives. More somberly, they are also the generation that entered the workforce just as the jobs were dying. And so they have splintered into two clear and predictable tiers: those who bob from job to job because they can and sense that they should, and those scrambling between multiple part-time gigs, none of which lays a path to prosperity as a union job once did.

Finally, the workplace of the future will be increasingly flattened and intermediated, deconstructed by digital technologies that twist both time and space. Already, things like mobile videoconferencing and remote file sharing allow workers to be working even when they're not at work. People can—and do—work from the car, on the subway, by the side of their kid's soccer game, or from the waiting room at the dentist. They can work anytime, all the time, in any time zone they have flown through or might fancy. And in return, they now have a million industrious-looking ways not to be working at work—see Instagram, Candy Crush, and pretty much every other app on a mobile phone. The lines between work time and free time, between what used to be labor and what now counts as fun, are becoming increasingly blurred, migrating to the same devices and speaking the same code. In some ways, the work world of the future could resemble that of the preindustrial past, with smaller groups of people interacting for one-off transactions, melding business and pleasure across a series of shifting, overlapping, technology-mediated networks.

This changing face of labor, though, means that we will need, as a society, to find something to *do*, some way for people to find meaning in a world increasingly shorn of work. Both economically and existentially, we will need to figure out what it means to be productive as our lives grow longer and machines take care of what were once subsistence needs.

We will need to imagine, as Keynes foresaw, "how to occupy [our] leisure . . . [how] to live wisely and agreeably and well."

Doing so will not be easy, especially given the inequities implied by
a shrinking working class. But these are still fixable problems, made
more addressable by an acknowledgment of what is driving them: technology is rendering labor increasingly obsolete, shattering the kinds of
jobs that shaped both agricultural and industrial societies, and forcing
us to wrestle with what will become a central issue of our postindustrial
future: What do we do as a society when all the jobs are dead?

The Future of Sex

To understand where sex is likely to go, we need to start with its
urgent and messy by-product: babies.

Ever since the pill made contraception pretty close to perfect, it's
been easy for all of us—men and women, married and single—to forget
how tightly sex and pregnancy once were linked. For a young, fertile
woman who is having sex regularly, the chance of becoming pregnant
each month is 50 percent. *Fifty percent.* Throughout nearly all of history,
therefore, from way before the Neolithic Age until roughly 1960,
women were pregnant, or about to be pregnant, or trying desperately
not to become pregnant, or trying desperately *to* become pregnant *all
the time.* Pregnancy was essentially what defined women for thousands
and thousands of years, and pregnancy was inextricably linked to sex.
As Firestone fumed in the *Dialectic*: *"Pregnancy is barbaric."*[10]

When the pill came along (together with safer IUDs, better dia-
phragms, and legal access to abortion), this link was severed. Forever. In
just the tiniest blink of historical time, sex was sprung from its biological
purpose and women were free, for the very first time, to separate their
sexual selves from their procreative ones. The technology of contraception created sexual liberation, enabling women (and men) to forever
after mix and match their romantic desires and parental ones. Whereas
pregnancy in the past had been a virtually certain result of marital sex,

contraception made conception a choice. In the post-pill world, individuals could enjoy sex without marriage and marriage without pregnancy; they could develop sexualities that weren't pinned to an overarching structure of man-woman-child.

And so they did.

Specifically, in the decades that followed the pill's release, sex outside marriage became a norm instead of a sin, and even committed couples usually engaged in sex long before marriage. Sexuality itself became something to experiment with, to play with, to enjoy and desire without either the fear of pregnancy or the pressure to produce a child. Long-forbidden sexual preferences could flourish in a world that didn't tie sex so tightly to marriage; sexual identities could, and did, become more fluid.

Then, just as these changes were settling across the once-stable configurations of family life, technology intervened again, bringing both the rise of assisted reproduction and the introduction of online romance. Typically, these are treated as wholly separate spheres, one related to how married people produce children, the other to how they get married in the first place. But they are actually deeply related, since both are tugging—pulling, yanking—at the underlying structures of sex and marriage.

Whereas assisted reproduction was once used to maintain a facade of normalcy in marriages that could not otherwise produce a child, it has now burst across an ever-wider range of parents, and partnerships, and desires. Gay couples are using—must use—the technologies of assisted reproduction to produce a child. Single people—both men and women—are relying on surrogates, sperm donors, and IVF to become parents. Soon, technology may allow wholly platonic threesomes, or elderly women, or combinations we can't even yet imagine to become parents. At that point, the link between sex and procreation, much less romance and procreation, will be fundamentally severed. In fact, it probably already is. Which means that sex becomes—well, just sex. It can be casual sex, uncommitted sex, same-sex sex, virtual sex, adulterous sex, or sex with a chatbot who knows your name. It can be all of these things at once (almost), or a variety of them sampled over the course of one's life. Driven by technologies of conception and contraception, sex, like work, becomes much more amorphous and free-flowing, less confined by major social structures and driven instead by an individual's choice, and chance, and whimsy.

And this is precisely the kind of sex that prevails and predominates online. Sex on demand. Sex shorn of commitment. Sex (in theory at least) with such a wide array of possible partners that it starts to feel like a game. "It's like a boredom cure," said one young man to me recently, describing his bouts with Tinder. "When you get a match, it feels like bingo."[11] Not exactly a hearts-aflutter kind of thing. But it mirrors the style of a generation raised on gaming and accustomed to choice, to swiping left and swiping right across an endless parade of options. By contrast, consider the recurring theme of *When Harry Met Sally*, Nora Ephron's 1986 exploration of the power and perils of sex. As they meet over the years and through various other relationships, the two protagonists remind themselves, "Men and women can't be friends, because the sex part always gets in the way." And Harry and Sally can't. After having sex, and then pulling apart, they reunite on New Year's Eve, to be happily married thereafter. For my bingo-playing student, by comparison, the sex really *doesn't* get in the way—because it happens without any pretense, and far before the question of friendship, much less love, has time to interfere.

In retrospect, the moment of Harry and Sally was a liminal time, poised almost perfectly between the sexual liberation of the 1960s and the online explosion to come. It was a time when sex had been decoupled from marriage and procreation, but not from love. Peering into the future, though, it seems almost certain that all three of these links will continue to fray, with both positive and negative implications for the future of sex. On the positive side, the vanishing connection between sex and reproduction will, as Firestone predicted, finally free women from the physical and emotional burdens of pregnancy. Not immediately, of course, and not entirely: until the creation of artificial wombs (theoretically plausible but still far away), women will still be the ones to carry, birth, and nurse their children. Their bodies will bear the emotional uncertainties of conception and the physical strains of pregnancy. But as reproduction increasingly becomes something planned and programmed, as couples sit down with genetic counselors and menstrual trackers rather than oysters and champagne, the burdens that once fell wholly on women—getting pregnant, not getting pregnant, keeping track of whether or not they might be pregnant—will slowly but surely ease. Women will be able to have sex without babies, babies without sex, and both sex and babies without men—all of which should free them to make the choices they want, both personally and professionally.

Both men and women, meanwhile, will also be free to play a more conscious role in their children's conception, and to experiment with sexual arrangements and identities that once fell outside the pale. If the world of the past was a Noah's Ark pathway of rigid roles and rules, the future will be considerably more free. Polyamory. Bisexuality. Threesomes. More-somes. Sex with silicone dolls and ever more realistic avatars—it's all out there and on your phone.[12] Sexual preferences that have long been deemed either taboo or marginal are likely to become more commonplace—a realization, perhaps, of Marcuse's "erotic reality" or what Firestone called humanity's "natural polymorphous sexuality."[13] Every so often, during one of my small focus groups, a twenty-something would turn to me pointedly and ask, "Don't people of your generation regret not having sex the way ours does?"

Meanwhile, the newfound ability for trans men and women to assume their preferred gender has also been truly revolutionary—not only for the people directly involved but for the next generations of society as well. Because once gender itself becomes elastic and transformable, once the most basic organization of society can be adapted to an individual's most intimate desire, every other social hierarchy is likewise burst open for inspection. No wonder that archconservatives react so strongly: once the binary is gone, there's not much left to hold the future back.

On the negative side, however, the shock of the sexual new has undeniably left some reeling in its wake. Not so much in the reproductive realm, where new technologies tend swiftly to defeat any initial resistance that surrounds them. Or even in the area of gender transition, where opposition will hopefully prove churlish and short-lived. No, the real downside of these technology-induced changes has been in the sheer infinity of options they have created for some people, the buffet of choices that can, after a while, feel overwhelming and lonely and sad. This isn't the part of the techno-sexual revolution that cries out for commentary. It's not about the titillation of virtual sex or anonymous hookups. Instead, it's about how infrequent these hookups actually are for many people, and how lonely such interactions can be. In theory, the constantly connected technologies of the internet and social media allow people all over the world to find each other, attract each other, and indulge in whatever guilt-free sexual adventures their hearts or loins desire. And some— many—do. But technological change always leaves some behind, and technologies for love or lust are no different. There are winners in the new world of sex and losers; those cut free from historical constraints and the bad luck of biology, and those who might have been better off in a time of fewer options.

As these technologies evolve, then, it will be crucial to ensure that the chasm between the inevitable haves and have-nots does not loom too large. We don't want a world that clusters reproductive and sexual favors on the rich and beautiful, any more than one that reserves all the good jobs for a favored few. We don't want a world where the gamification of sex destroys the joy it can bring. But working toward that more symmetrical world means acknowledging, at the very outset, the web of connections between sex and technology, and the extent to which our most intimate lives are being shaped by distant inventions.

The Future of Marriage

What, then, of marriage, that happily-ever-after state that has dominated our collective fantasy for the past several hundred years?

Well, the first thing to keep in mind is that marriage is a decidedly *social* institution. And thus as society evolves in response to technological change, so will marriage.

Already, in fact, these changes are well underway. Between 1960 and 2016, the crude marriage rate (defined as the number of marriages per one thousand people) dropped from 7.8 to 3.7 in Spain. In Germany, likewise, it fell from 9.5 to 5.0; in the Netherlands, from 7.7 to 3.8.[14] In the United States, as noted in chapter 5, only 22 percent of men and 33 percent of women were married before turning twenty-five in 2016, down from 72 percent and 87 percent, respectively, in the early 1960s.[15] In Japan, the marriage rate plummeted nearly 50 percent over that same time period.[16]

Outside the sheer statistics, moreover, the shape of marriage is changing as well. Most dramatically, and despite opposition from conservative forces, marriage is no longer defined as occurring just between a man and a woman. In most parts of the world, men can now be married to other men, and women to other women. This is an extraordinary development—a wonderful, important, life-affirming development, as the advocates of same-sex marriage have long argued. But it is also, as their opponents have noted, a major change. Marriage as an institution *is* different once it expands to include same-sex partners, and this expansion will make it harder—both legally and culturally—not to accept other expansions and redefinitions in the future. Interracial and interethnic marriages are on the rise in the United States and elsewhere;[17] couples like Martine and Bina are staying married even after one of the partners changes gender and what was once a heterosexual marriage becomes technically a homosexual one. In Australia, the courts in 2011 blessed the legal union of three people.[18]

In reviewing these recent changes, it's useful to remember the conditions under which marriage initially arose and was cemented. Hatched in the cradle of the first agricultural revolution, it was an institution that enabled children—heirs—to be conceived, born, and cared for. It gave men the certainty of paternity, and women the promise of financial support. It emerged at a time when contraception was ineffective, sexual intercourse was the only way of producing a child, and most people died before forty. An awful lot has changed since then.

In particular—and this is so basic that it bears repeating—we no longer, as a species, need sex to make babies. Which means, in turn, that we no longer need the institution of marriage to sanctify and secure the conditions under which our offspring are born. We have DNA tests instead to determine paternity, and donor eggs, sperm, and wombs to help fashion a child. We can select the genes we desire to carry our legacy, and can tweak them even further, should we need to, at the embryonic stage. Our ancient ancestors were lucky if they lived long enough to see their children wed. We live now, as a species, long past our children's maturity, long past what was once our reproductive prime. Long past the first pangs of love. Societally, we simply don't need marriage as we once did. Sex—yes, we are biologically programmed to want that. Work—we seem programmed to want that, too, although there are clearly many ways to productively spend one's time other than laboring for subsistence. But marriage? We created it in the relatively recent past, building it upon a societal foundation of familial obligation, religious observation, and law. And, as our needs and technologies change, we can dismantle and rearrange it again.

Indeed, I suspect we shall.

The Future of Love

Finally, and logically, as sex and work and families evolve, so will love. Not the emotion itself, one hopes, or the burning-heart, broken-heart, hopeful-heart passion it evokes. But the form it takes, perhaps, and ways in which it is manifested. One hundred years ago—even fifty—a Jew couldn't safely love a Catholic. It wasn't condoned, certainly, and social structures would have separated the would-be pair. A black man couldn't safely love a white woman. A man couldn't safely love another man. Now these combinations are not only possible, but commonplace. The most basic love—the love of parents for their children—is evolving as well, expanding to include offspring acquired through an ever-widening array of channels. If we hit the science fiction heights that are becoming seductively possible, love could even grow immortal, captured in the digitally stored consciousness of a lover's life, or restored in the form of prosthetic bodies and artificially intelligent minds.

Indeed, sketching out the landscape of technological change, it's intriguing to consider just how large a role love has played—not in driving the mechanics of invention per se, but in shaping the needs and desires that the fruits of innovation are being created to serve. Consider the field of assisted reproduction. Over the past century, scientists working in this area have made massive, pathbreaking discoveries: synthesizing hormones, engineering in vitro conception, unlocking the patterns of embryonic genetic development. The entire field of stem cell medicine had its roots in assisted reproduction as well, since it was donor eggs and excess embryos that provided stem cell scientists with the tools and raw material for their early work.[19] And what drove this work? Not the crazy desire of a megalomaniac to clone a thousand copies of himself, or a dictator's plan to generate armies of genetically engineered soldiers. No, it was the mundane, ordinary, beautiful desire of millions of men and women to create something—someone—to love.

So, too, with dating apps and assistive robots. These are massive, complicated, system-wrenching technologies, yet their central purpose is to help individuals find and tend to love. Yes, many people use Tinder and its peers to find lust before love; and yes, one can see robots like Pepper and Paro as poor substitutes for human care and affection. But in my admittedly brief forays into both these worlds, what struck me was the inherently old-fashioned purpose behind their use. The elderly nursing home residents aren't cuddling with Paro because they admire his space-age technologies or beguiling algorithm; they are responding to a non-human being with a very human form of love, or at least affection. And for every bingo man scrolling Tinder for sex or amusement, there are lots of other lonely users looking instead for a human attachment, for something, even something fleeting or momentary, that looks and feels like love. I wouldn't go so far as to say that love is a causal driver of technological change. Profits, probably, are a bigger force, along with humankind's ancient and unvarnished desire to know more, do more, build more—to go, always, where no one has gone before. But whereas social constructions such as work and marriage are liable to be fundamentally uprooted by technology's next wave, love seems likely to remain unscathed. Or at least changed in ways that don't erase its underlying pull and power.

Some of these possible iterations are undeniably scary. We recoil—most of us, at least—at the idea of falling in love with a shapeless avatar, or of being comforted in our old age by a robotic seal. We feel a certain uneasiness hearing about individuals who engineer their bodies to hover between genders or those who have already taken steps to freeze their brains. But we have felt this fear before—and then promptly gotten over it. Remember the dangerous, smoke-belching car? The diaphragm? Both were roundly denounced at the moment of their invention. Remember the uproar that surrounded Louise Brown, the first test-tube baby? Well, Louise is an ordinary middle-aged woman now, and the furor that arose over in vitro fertilization has long since died down. This same gradual acceptance is bound to occur for whatever new and crazy things we do in pursuit of love. If I can fall in love with a random stranger across the room, what's so far-fetched about loving a robotic seal or trying to preserve your lover's mind?

Cynthia Breazeal, director of the Personal Robots Group at the MIT Media Lab, believes that people respond to machines just as they do to other people. "We are profoundly social and emotional beings," she argues. "We crave interaction . . . That is what it means to be human."[20]

And Anders Sandberg, a research fellow at Oxford's Future of Humanity Institute, sees no reason why couples in love shouldn't already be planning for their joint cryo-preservation. He wears a medallion around his neck, etched with wedding vows made recently to his husband: "Till death do us part . . . temporarily."[21]

Indeed, in visiting with some of these pioneers and their technological creations, I was repeatedly struck not by the novelty of what they're trying to do, but rather with the stodgy old-fashionedness of it. Martine Rothblatt is trying to make a great love stay alive forever. Takanori Shibata wants old people to feel joy. Ray Kurzweil wants to talk with his father again. And the trans students I came to know at Barnard and elsewhere were simply trying to get on with their lives, to love themselves and their bodies in ways that most people have the luxury of taking for granted.

It seems strange to believe that we could undergo a seismic change as a species and come out still lovelorn on the other side. But that's where we seem, in these very early days of the revolution, to be heading. We are changing the face of labor, and the relative position of men and women across the workforce and home front. We are having sex more impetuously, with less commitment and in an expanding range of socially acceptable permutations. We are deconstructing the institution of marriage, paving the way for a future in which the nuclear family, itself a creation of the Industrial Revolution, will decline in prominence and import. But we still yearn to love, and to find those—online, at work, created via test tube or algorithm or reanimated brain—who will learn to love us back.

Re-Conceptions

If the future of the past was a bright and shiny thing, then the past of our collective future is still a work in progress. We can see that future coming, barreling toward us, but we don't yet know precisely how it— and we—will evolve.

The first thing to note, though, is that our ability to stop this evolution is severely limited. Over and over, since we first shimmied down from the treetops to learn to hunt and farm, our human species has been making things—from axes and plows to steam engines, automobiles, and smartphones. Indeed, it may be tool-making—technology— that defines us as a species and that has propelled our development until this point. More critically, when we make things, we use them: we farm with plows, power railroads with steam, and conduct our lives' business now on our phones. Over thousands and thousands of years, we have almost never resisted technology's advance or refused to use the tools we have built. Instead, after a predictable period of moaning and groaning, of decrying the pace of change and mourning for the past, we embrace our new machines and rebuild our lives around them. How many people really miss the world of rotary phones? How many people under the age of twenty-five even know, or care, what they were?

To be sure, some of the technologies of the impending digital age are sufficiently threatening to have already generated a wave of proactive opposition. Industry pioneers such as Elon Musk and Bill Gates, for example, have spoken publicly about the risks inherent in artificially intelligent machines.[22] Leading scientists such as Jennifer Doudna, who created the revolutionary CRISPR technology for genomic engineering, have argued that the technology should not be used to modify human embryos or germline cells until appropriate safeguards are in place.[23] Such fears are eminently well-founded. Because we truly don't know what artificial intelligence will prove capable of being or doing, or how genetic modifications might shape or even mutate our species over time. What history clearly teaches us, though, is that outright bans on these technologies are highly unlikely to succeed. (In fact, the only breakthrough technology that has been banned to any real effect is nuclear weapons, and even there one can argue that the weapons themselves have been built and deployed for what was always their stated purpose—deterrence.)[24] Instead of engaging in Luddite fantasies, therefore, the only way forward is through the messier, murkier process of regulation, carefully deploying guidelines around these emerging tools, and then adapting them as both society and technology change. I will return to this point below.

In the meantime, though, we are facing another threat, one less existential but more immediate. And that is the deeply asymmetrical way in which digital technologies are likely to be used, and the social inequities that are almost certain to be perpetuated as a result. Again, we've seen pieces of this pattern before: wealthy people had access to automobiles way before the working class did; automatic looms made a handful of industrialists into millionaires while consigning millions more to a life of factory drudgery. But digital technologies are liable to land with particular force, driven by their ever-expanding ability to replace human labor with smart machines. Truck drivers. Assembly-line workers. Tax preparers and bank tellers.[25] All are likely, within decades, to be gone. And the jobs that emerge in their wake will be sharply bifurcated, allowing small groups of people to generate massive wealth while others are mired in poorly paid service jobs and the piecework of the gig economy. Meanwhile, technologies for life extension and body hacking will give the wealthy unprecedented options for living not only better lives but longer and healthier ones as well. Put these together with the sexual asymmetries of online dating and the reproductive asymmetries of high-tech fertility treatments, and we start to approach a society of dizzying inequities. Indeed, without concerted intervention, our postindustrial future could drag us back to social structures that resemble the feudal past, to a world of well-fed, well-sexed lords of the manor reigning over a patchwork kingdom of Uber drivers and fast-food clerks. Just listen to the protests that arose across France in 2018, to the fight over Brexit in the United Kingdom and around the U.S. election of Donald Trump. People are angry. People are hurt. Their social norms and structures have been upended, and the glories of a technologically enhanced future are not necessarily theirs to reap. Yes, YouTube and Instagram can catapult a few lucky users to stardom; yes, Tinder and Hinge put thousands of potential partners in everyone's reach. But the economy of the future isn't pro-grammed to be fair. It never is.

To address these inequities, therefore, we need to start putting pol-icies in place: not bans on technology, which never work, but careful, cautious plans for modulating some of the digital economy's harshest edge. This isn't as hard as it may sound, especially once you (a) pull back from trying to regulate what doesn't yet exist; and (b) focus on the so-cial impact of technology rather than the technology itself. As the In-dustrial Revolution was unfolding across England, for example, the British government didn't ban automatic looms or railroads. Instead—slowly, to be sure, and often contentiously—it smoothed the social impact of innovation with things like speed limits and bans on child la-bor. Similarly, as techniques for assisted reproduction have become both more sophisticated and more commonplace, governments across the world (with the United States being a notable exception) have grad-ually built regulatory systems to address the waves of change. In Den-mark, for example, the state provides all women with access to three cycles of IVF regardless of their marital status or sexual identification.[26] In Israel, all would-be mothers between the ages of eighteen and forty-five are entitled to state-funded fertility treatments up until the birth of two live children.[27] The point isn't to get each policy exactly right, but rather to create a framework through which a society can adapt to tech-nological change, instead of simply being trampled by it.

Having a framework for technology review would also allow us to slide more cautiously down an admittedly slippery slope. When we are faced with potentially revolutionary change, our instinct all too often is to declare either resistance or retreat: we must fight the robotic forces or submit already to their demands. Ban all forms of genetic engineering or prepare for the world of the Übermensch. Embrace a world of seventy-one possible genders or insist on just two.[28] Yet reality is more nuanced than that, especially when it's in the process of being nudged and reshaped by technological change. We can draw lines across even shifting sands, and we must.

To be sure, many of technology's most intimate effects will always fall far beyond the political realm. We can't legislate against people's online loneliness, for example, or create a reallocation system for sex.[29] We can't tell people whom to fall in love with, or how. But as the histories of same-sex marriage and assisted reproduction demonstrate, we can and do allow laws to shift over time, evolving alongside both technological and social change. Indeed, if technological innovation drives social change, as these pages have argued, then we have a powerful lens through which to see not precisely how that change will unfold, since the future is bound to trick us, but at least its general tilt and direction, so that we can anticipate its most important effects.

Some of these are straightforward. Over the next few decades, we are liable to experience a massive loss of jobs across the industrial economy, and an accompanying rise of inequality. Men will be particularly hard-hit by these shifts, as will manufacturing workers more generally, and those without a college education. We must start preparing for these shifts—ideally, through a combination of vocational training, educational partnerships, and income-redistribution options that could range anywhere from educational tax credits to a universal basic income.[30] We need to encourage more open conversations about male identity and masculinity, and to support men who are carving out new roles as caregivers instead of breadwinners.

Similarly, we know that families are bound to evolve—perhaps quite dramatically—in the not-too-distant future. As sex and reproduction become ever less connected, as sex becomes gamified and genders blur, the traditional family of the industrial age will become increasingly irrelevant for growing segments of the population. And yet that fast-fading archetype is still the model for most of our policies and much of our workaday lives. The entire U.S. income tax system, for instance, is stuck in the 1950s, presuming (and favoring) a family structure composed of a male breadwinner, his female spouse, and a few dependent children. Surely, that can change? As can the less formal but equally rigid norms that prevail at schools and workplaces. Why assume that workers have a spouse, or even a quasi-permanent partner? Why assign benefits along those lines, or restrict family leave to dealing with children or elderly parents? And why must public schools have increasingly high expectations for parental involvement—just as more and more parents are either working, or single, or both? All these contradictions, caught between an aging social structure and a new one, should be relatively easy to address.

So, too, with assisted reproduction, despite the bouts of hand-wringing that so frequently surround it. People are going to have babies in different ways, in different configurations, and at different moments in their lives. They are doing so already. At some point, they are also going to begin engineering their offspring more consciously than is now possible. We need to prepare for these eventualities—not by banning any specific technology, but rather by constructing a careful framework of regulation, one that can consider these technologies as they emerge and find ways to best guide and pay for them. As noted above, many governments already have such systems. Others should be watching, and following suit.

Other changes, though, are admittedly harder to predict and prepare for. We know that the robots are coming, for instance, and that they, along with all our digitally driven machines, are getting smarter and smarter. We know they will disrupt our jobs and factories; I think it's also fair to predict that at least some of us will fall in love with some of them. But matters after that get murky—both because the future is inherently hard to envision and because technologies such as deep learning, brain mapping, and mind uploading remain in their relative infancy.

In the course of writing this book, though, I have personally become convinced that futurists like Ray Kurzweil and Martine Rothblatt are right: that at some point in the not-so-distant future, we will merge with our machines, becoming in the process something both more human and less. It may be the bit-by-bit process that some body hackers foresee, adding a bionic arm here and an artificial eye there. Or the all-at-once full-brain conversion that some neuroscientists are starting to imagine. It could be that we will make our machines so smart that they plot to replace us, or that we will, over time, just need them more and more. What sounds so scary at the outset—*The machines! Taking over!*—is in many ways only the updated version of the story humankind has been telling itself since our storytelling days began. We are a tool-making species, and a questioning one, probing to understand both the source of our life and its meaning.

Technology is a tricky thing. It is our creation, a wholly human construction, yet developed now to a point where it could—might, probably will—expand beyond our human ability either to comprehend its inner workings or to completely control them. We have built the pieces—millions of them, crafted by generations of us—without a master plan for how these pieces should eventually interact and evolve.

And thus we face our future with a strange admixture of reverence and fear. Reverence for the power we have mustered and its looming ability to free us from the bodily constraints that have bound us for so long. And fear that we may have stumbled into something wrong. Something dangerous and final, something that could extinguish the will we see as being distinctly ours. It would be foolish to predict how this tension will ultimately be resolved, or how our machines will morph and coexist with us. Like our hunting-and-gathering ancestors who moved slowly toward farming, or our farming ancestors who saw the first railroad racing across the horizon, we are looking toward a future whose outlines we can only barely discern. And nothing, as *Frankenstein*'s narrator solemnly intoned in 1818, "is so painful to the human mind as great and sudden change."[31]

NO ONE TODAY can foretell how our next-generation descendants will emerge, or how we will adapt to interact with them. Yet as we consider our creations, it is crucial to remember that the machines that we have fashioned have no goals beyond our own. No lives, no loves, no passions. Their dreams are whatever we impose upon them. Therefore, as we hit the inflection point of the revolution we have wrought, we need to wrestle explicitly with what we want them to be, and how we want to live as humans among them.

As long ago as ancient Rome, we began to dream of creating our own descendants, of building machines that would become like us and allow us to be better.

Now, at last, the technology is here.

And it's our turn to evolve.

Notes

Prologue: The Futures of the Past

1. Quotes are from a contemporary description, quoted in Peter Berlyn, *A Popular Narrative of the Origin, History, Progress, and Prospects of the Great Industrial Exhibition, 1851* (London: James Gilbert, 1851), p. 140.
2. Jeffrey Stanton, "Showcasing Technology at the 1964–1965 New York World's Fair," *New York 1964 World's Fair*, at https://www.westland.net/ny64fair/map-docs/technology.htm.
3. Lisa L. Colangelo, "1964 World's Fair: When the World Came to Queens," *New York Daily News*, April 2014, at http://creative.nydailynews.com/worldsfair.

4. See, for example, James Fenske, "African Polygamy: Past and Present," *Journal of Development Economics*, vol. 117 (November 2015), pp. 58–73; Alberto Alesina, Paula Guiliano, and Nathan Nunn, "On the Origins of Gender Roles: Women and the Plough," *Quarterly Journal of Economics*, vol. 128, no. 2 (May 2013), pp. 469–530; and Jack Goody and Joan Buckley, "Inheritance and Women's Labour in Africa," *Africa*, vol. 43, no. 2 (April 1973), pp. 108–21.

5. See, in particular, "Preface to *A Contribution to the Critique of Political Economy*," "The German Ideology," and "The Poverty of Philosophy," all in Robert C. Tucker, *The Marx-Engels Reader*, 2nd ed. (New York: W. W. Norton, 1978). For a concise overview, see Peter Singer, *Marx: A Very Short Introduction* (Oxford: Oxford University Press, 2018).

1. Life Before the Machines

1. Cited in Charles C. Mann, *1491: New Revelations of the Americas Before Columbus*, 2nd ed. (New York: Vintage, 2011), p. 20.

2. The dates are uncertain here, and themselves keep evolving along with archaeological discoveries and improved techniques for assessing age. Five hundred thousand years ago is the date currently estimated for when Neanderthals, an early human species, evolved in Europe and the Middle East. See Yuval Noah Harari, *Sapiens: A Brief History of Humankind* (New York: HarperCollins, 2015), p. viii. For a discussion (and later estimate of their arrival), see Richard E. Green et al., "Analysis of a Million Base Pairs of Neanderthal DNA," *Nature*, vol. 444, no. 7117 (November 16, 2006), pp. 330–36. For a general discussion, see Clive Finlayson, *The Humans Who Went Extinct: Why Neanderthals Died Out and We Survived* (Oxford: Oxford University Press, 2009).

3. Again, the precise dates are subject to revision. I am following the timeline set out in Harari, *Sapiens*, pp. viii, 3–19. For broad overviews of this long period, see also Tim Megarry, *Society in Prehistory: The Origins of Human Culture* (New York: New York University Press, 1995); and Kaye E. Reed, John G. Fleagle, and Richard E. Leakey, *The Paleobiology of Australopithecus* (New York: Springer, 2013).

4. Harari, *Sapiens*, pp. 48–53; T. Douglas Price and James A. Brown, eds., *Prehistoric Hunter-Gatherers: The Emergence of Cultural Complexity* (Orlando: Academic Press, 1985); Geoff Bailey, ed., *Hunter-Gatherer Economy in Prehistory: A European Perspective* (Cambridge, UK: Cambridge University Press, 1983); and Emma Groeneveld, "Prehistoric Hunter-Gatherer Societies," *Ancient History Encyclopedia*, at https://www.ancient.eu/article/991/.

5. For a review of these schools of thought, see Stephanie Coontz, *Marriage, a History: From Obedience to Intimacy or How Love Conquered Marriage* (New York: Penguin Books, 2006), pp. 37–38. See also R. Lee and I. DeVore, eds., *Man the Hunter* (Chicago: Aldine, 1968); and Tim Ingold, David Riches, and James Woodburn, eds., *Hunters and Gatherers 1: History, Evolution, and Social Change* (Oxford: Berg, 1988).

6. Or, as Harari puts it, "social cooperation is our key for survival and reproduction." Again, the dates are complicated. I am following Harari, who puts the evolution of *Homo sapiens* at roughly two hundred thousand years ago, and the emergence of

Homo sapiens as the dominant species at around seventy thousand years ago. Harari also defines this latter period as the start of humankind's cognitive revolution. See Harari, *Sapiens*, pp. viii–40; the quote is from p. 22. The role of social groups also features prominently in anthropological accounts of other preindustrial societies. See, for example, Margaret Mead, *Coming of Age in Samoa* (New York: William Morrow Paperbacks, 2001); Kim R. Hall et al., "Co-Residence Patterns in Hunter-Gatherer Societies Show Unique Human Social Structure," *Science*, vol. 331, no. 6022 (March 11, 2011), pp. 1286–89; and Coren L. Apicella et al., "Social Networks and Cooperation in Hunter-Gatherers," *Nature*, vol. 481 (January 26, 2012), pp. 497–501.

7. See Harari, *Sapiens*, pp. 45–53; and Groeneveld, "Prehistoric Hunter-Gatherer Societies."

8. Early anthropologists believed that preindustrial people were distinctly polygamous, engaging in what Lewis Henry Morgan, a nineteenth-century anthropologist who spent decades observing the Iroquois and other Native American tribes, described as a state of "promiscuous intercourse." Margaret Mead, writing in the 1920s, famously recounted the sexual freedoms enjoyed by adolescent girls in the Samoan Islands, stating, for example, that "Samoans rate romantic fidelity in terms of days or weeks at most, and are inclined to scoff at tales of life-long devotion." More recent work, however, suggests that polygamy was less prevalent, and that "pair-bonding," as Helen Fisher argues, "is a trademark of the human animal." Along with this bonding, though, Fisher also cites the widespread prevalence of adultery among preindustrial groups. See L. H. Morgan, *Ancient Society or Researches in the Lines of Human Progress from Savagery Through Barbarism to Civilization* (Chicago: Charles H. Kerr, 1877/1908); Margaret Mead, *Coming of Age in Samoa*, p. 108; and Helen Fisher, *Anatomy of Love: A Natural History of Mating, Marriage, and Why We Stray*, rev. ed. (New York: W. W. Norton, 2016), pp. 59–80. Other accounts of early sexuality include C. S. Ford and F. A. Beach, *Patterns of Sexual Behavior* (New York: Harper and Brothers, 1951); P. L. van den Berghe, *Human Family Systems: An Evolutionary View* (Westport, CT: Greenwood Press, 1979); and, more controversially, Christopher Ryan and Cacilda Jethá, *Sex at Dawn: The Prehistoric Origins of Modern Sexuality* (New York: HarperCollins, 2010).

9. Megarry, *Society in Prehistory*, p. 292.

10. Friedrich Engels, *The Origin of the Family, Private Property and the State* (New York: International, 1942).

11. See, for example, M. Dyble et al., "Sex Equality Can Explain the Unique Social Structure of Hunter-Gatherer Bands," *Science*, vol. 348, no. 6236 (May 14, 2015), pp. 796–98; Michael Gurven, "To Give or Not to Give: An Evolutionary Ecology of Human Food Transfers," *Behavioral and Brain Sciences*, vol. 27, no. 4 (2004), 543–83; Paula K. Ivey, "Cooperative Reproduction in Ituri Forest Hunter-Gatherers: Who Cares for Efe Infants," *Current Anthropology*, vol. 41, no. 5 (December 2000), 856–66; and Courtney L. Meehan et al., "Cooperative Breeding and Maternal Energy Expenditure Among Aka Foragers," *American Journal of Human Biology*, vol. 25, no. 1 (January 2013), 42–57. More generally, see also Margaret Mead, *Male*

and Female: A Study of the Sexes in a Changing World (New York: William Morrow, 1949), esp. pp. 189–90.

12. See Lynn White Jr., *Medieval Technology and Social Change* (New York: Oxford University Press, 1966), p. 41.

13. For a seminal study on the long-term implications of plow-based agriculture, see Alberto Alesina, Paola Giuliano, and Nathan Nunn, "On the Origins of Gender Roles: Women and the Plough," *Quarterly Journal of Economics*, vol. 128, no. 2 (May 2013), pp. 469–530. For an argument that women were likely responsible for the development of the hoe, and the earliest forms of farming, see Hermann Baumann, "The Division of Work According to Sex in African Hoe Culture," *Africa: Journal of the International African Institute*, vol. 1, no. 3 (July 1928), pp. 298–319.

14. Harari, *Sapiens*, p. 77; James C. Scott, *Against the Grain: A Deep History of the Earliest States* (New Haven: Yale University Press, 2017), pp. 1–35.

15. Douglass C. North and Robert Paul Thomas, "The First Economic Revolution," *Economic History Review*, vol. 30, no. 2 (1977), p. 235.

16. For an interesting albeit controversial account that instead describes private property as an outgrowth of power, see David Graeber, *Debt: The First 5,000 Years* (Brooklyn: Melville House, 2011).

17. For more on this shift, see Robert S. McElvaine, *Eve's Seed: Biology, the Sexes, and the Course of History* (New York: McGraw-Hill, 2001), pp. 87–91.

18. See Alesina, Giuliano, and Nunn, "On the Origins of Gender Roles: Women and the Plough," pp. 469–530. The original work for this hypothesis comes from Ester Boserup, *Woman's Role in Economic Development* (London: George Allen and Unwin, 1970).

19. These estimates come from both archaeological analyses of fossils and the recorded practices of modern-day hunter-gatherer societies. For a review of the evidence and estimates, see Casper Worm Hansen, Peter Sandholt Jensen, and Christian Volmar Skovsgaard, "Modern Gender Roles and Agricultural History: The Neolithic Inheritance," *Journal of Economic Growth*, vol. 20, no. 4 (December 2015), pp. 365–404.

20. Fisher, *Anatomy of Love*, p. 285.

21. The estimate comes from Richard B. Lee's work in the 1950s and '60s among the !Kung Bushmen of Botswana, nomadic hunter-gatherers who were then in the process of transitioning to a more sedentary life. See Richard B. Lee, "Population Growth and the Beginnings of Sedentary Life Among the !Kung Bushmen," in Brian Spooner, ed., *Population Growth: Anthropological Implications* (Cambridge, MA: MIT Press, 1972), p. 331. For more on the evolutionary effects of child-carrying, see J. C. Watson et al., "The Energetic Costs of Load-Carrying and the Evolution of Bipedalism," *Journal of Human Evolution*, vol. 54, no. 5 (June 2008), pp. 675–83.

22. See Scott, *Against the Grain*, pp. 113–15; and Lee, "Population Growth and the Beginnings of Sedentary Life Among the !Kung Bushmen," pp. 329–42.

23. See Luis Locay, "From Hunting and Gathering to Agriculture," *Economic Development and Cultural Change*, vol. 37, no. 4 (July 1989), pp. 737–56; and Claude Meillassoux, *Maidens, Meal and Money: Capitalism and the Domestic Economy* (Cambridge,

UK: Cambridge University Press, 1981). For a contemporary ethnographic account, see Lee, "Population Growth and the Beginnings of Sedentary Life Among the !Kung Bushmen," pp. 329–42.

24. See, for example, Scott, *Against the Grain*, pp. 150–82; and I. M. Diakanoff, *Structure of Society and State in Early Dynastic Sumer*, Monographs of the Ancient Near East, vol. 1, no. 3 (Los Angeles: Undena, 1974).

25. There is a well-documented link between fertility and the advent of a society's agricultural revolution. Simply put, as societies embrace the technologies of the Neolithic Revolution, their birth rates increase. See Hansen, Jensen, and Skovsgaard, "Modern Gender Roles and Agricultural History," pp. 365–404; Quamrul Ashraf and Oded Galor, "Dynamics and Stagnation in the Malthusian Epoch," *American Economic Review*, vol. 101, no. 5 (2003), p. 41; and T. Iversen and F. Rosenbluth, *The Political Economy of Gender Inequality* (New Haven: Yale University Press, 2010), pp. 32–33.

26. Scott, *Against the Grain*, p. 181.

27. Helen E. Fisher, *Anatomy of Love: The Natural History of Monogamy, Adultery, and Divorce* (New York: W. W. Norton, 1992), p. 287. See also Fisher, *Anatomy of Love*, rev. ed., pp. 63–68.

28. Hansen, Jensen, and Skovsgaard, "Modern Gender Roles and Agricultural History," p. 400. See also L. Fortunato and M. Archetti, "Evolution of Monogamous Marriage by Maximization of Inclusive Fitness," *Journal of Evolutionary Biology*, vol. 23, no. 1 (2010), pp. 149–56.

29. Coontz, *Marriage, a History*, pp. 34–49.

30. See S. Greengus, "Old Babylonian Marriage Ceremonies and Rites," *Journal of Cuneiform Studies*, vol. 20, no. 2 (1966), pp. 55–72, at https://www.jstor.org/stable /1359030?seq=1, accessed November 2019; Elisabeth Meier Tetlow, *Women, Crime and Punishment in Ancient Law and Society*: vol. 1, *The Ancient Near East* (New York: Continuum, 2004), pp. 11–21, 26–27, 41–49, 75–76.

31. See Jacob J. Finkelstein, "Sex Offenses in Sumerian Laws," *Journal of the American Oriental Society*, vol. 86, no. 4 (October–December 1966), pp. 355–72; Raymond Westbrook, *Old Babylonian Marriage Law* (Horn, Austria: Verlag Ferdinand Berger & Söhne Gesellschaft M.B.H., 1988), pp. 29–38; and Russ VerSteeg, *Early Mesopotamian Law* (Durham, NC: Carolina Academic Press, 2000), pp. 78–84. For links between property and virginity more broadly, see Alice Schlegel, "Status, Property, and the Value on Virginity," *American Ethnologist*, vol. 18, no. 4 (November 1991), pp. 719–34.

32. Strict monogamy, with both partners pledging sexual fidelity, remained an exception until recently. See Walter Scheidel, "Monogamy and Polygyny," Princeton/Stanford Working Papers in Classics, January 2009, Stanford University.

33. "The 10 Largest Harems in History," FutureScopes, at http://futurescopes.com /polygamy/10266/10-largest-harems-history.

34. From the *Apastamba Dharmasūtra*, quoted in Coontz, *Marriage, a History*, p. 46.

35. See Coontz, *Marriage, a History*, p. 47; and Tikva Frymer-Kensky, "Virginity in the Bible," in Victor H. Matthews, Bernard M. Levinson, and Tikva Frymer-Kensky, eds., *Gender and Law in the Hebrew Bible and the Ancient Near East* (Sheffield: Sheffield Academic Press, 1998), pp. 79–96.

36. For this and all Sumer-related details, see E. J. Bickerman and Morton Smith, *The Ancient History of Western Civilization* (New York: Harper & Row, 1976), pp. 24–33. For population estimates, see Colin McEvedy and Richard Jones, *Atlas of World Population History* (New York: Facts on File, 1978).

37. Quoted in Tetlow, *Women, Crime and Punishment in Ancient Law and Society*, p. 9; and Marilyn French, *From Eve to Dawn: A History of Women*: vol. 1, *Origins* (New York: The Feminist Press, 2008), p. 100. Another document, called the Reforms of Uruinimgina, which dates from around 2350 B.C., reads, "The women of the former days used to take two husbands, but the women of today (if they attempt to do this) are stoned with the stones inscribed with their evil intent." Quoted in Tikva Frymer-Kensky, *In the Wake of the Goddesses* (New York: Free Press, 1992), p. 79.

38. A full translation of the code, by L. W. King, is available from the Avalon Project at Yale Law School, at http://avalon.law.yale.edu/ancient/hamcode.asp. The statutes referenced above are numbers 129 and 132. For more on the code and its social impact, see Benjamin R. Foster, "Social Reform in Mesopotamia," in K. D. Irani and Morris Silver, eds., *Social Justice in the Ancient World* (Westport: Greenwood Press, 1995), pp. 165–77; and V. L. Bullough, *Sexual Variance in Society and History* (Chicago: University of Chicago Press, 1976), pp. 51–53, 56–58.

39. French, *From Eve to Dawn*, p. 103; and David Herlihy, "The Triumph of Monogamy," *Journal of Interdisciplinary History*, vol. 25, no. 4 (Spring 1995), p. 578.

40. See Merlin Stone, *When God Was a Woman* (New York: Harcourt, Brace, 1976); Frymer-Kensky, *In the Wake of the Goddesses*; and E. O. James, *The Cult of the Mother Goddess* (London: Thames and Hudson, 1959).

41. See McElvaine, *Eve's Seed*, pp. 139–41; and Anne Baring and Jules Cashford, *The Myth of the Goddess: Evolution of an Image* (London: Penguin, 1991), pp. 145–74.

42. V. Gordon Childe, *Social Evolution* (New York: Henry Schuman, 1951), p. 65. In analyzing the creation stories of 112 societies, Peggy Reeves Sanday finds that increasing technological complexity encourages what she describes as a "masculine orientation." For example, while 46 percent of groups engaged in small-scale agriculture (fruit trees and gardening) have creation stories that are predominantly feminine, that number drops to 11 percent among groups practicing advanced agriculture. See Peggy Reeves Sanday, *Female Power and Male Dominance: On the Creations of Sexual Inequality* (Cambridge, UK: Cambridge University Press, 1981), Table 3.5, p. 69.

43. See Sanday, *Female Power and Male Dominance*, pp. 218–27. Deuteronomy is also explicit on this point: "You must completely destroy all the places where the nations you dispossess have served their gods," Yahweh commands. "On high mountains, on hills, under any spreading tree; you must tear down their altars, smash their pillars, cut down their sacred poles, set fire to the carved images of their gods and wipe out their name from that place" (Deut. 12:2–3).

44. See Raymond Westbrook, "The Female Slave," in Matthews, Levinson, and Frymer-Kensky, eds., *Gender and Law in the Hebrew Bible and the Ancient Near East*, pp. 214–38.

45. For a discussion, see Jacob L. Weisdorf, "From Foraging to Farming: Explaining the Neolithic Revolution," *Journal of Economic Surveys*, vol. 19, no. 4 (2005), pp. 561–86; and North and Thomas, "The First Economic Revolution," pp. 229–41.

46. For example, Hans J. Nissen, *The Early History of the Ancient Near East, 9000–2000 BC* (Chicago: University of Chicago Press, 1988); and Vernon L. Smith, "The Primitive Hunter Culture, Pleistocene Extinction, and the Rise of Agriculture," *Journal of Political Economy*, vol. 83, no. 4 (1975), pp. 727–55.

47. This is the central contention of James C. Scott's fascinating 2017 argument. See Scott, *Against the Grain*, pp. 1–35.

48. See Ester Boserup, *The Conditions of Agricultural Growth: The Economics of Agrarian Change Under Population Pressure* (Chicago: Aldine, 1965); and North and Thomas, "The First Economic Revolution," pp. 229–41. The data given here are from Scott, *Against the Grain*, p. 6.

49. See, for example, James E. McClellan and Harold Dorn, *Science and Technology in World History: An Introduction* (Baltimore: Johns Hopkins University Press, 2006), pp. 155–72.

50. For a description of the spread of agriculture and its implications in eastern and southern Africa, see Christopher Ehret, *An African Classical Age: Eastern and Southern Africa in World History, 1000 B.C. to A.D. 400* (Charlottesville: University Press of Virginia, 1998).

51. David Wu, "Great Yu: Mastering the Floods," *The Epoch Times*, October 16, 2012, p. B1.

52. For more on the connection between water and imperial power, see Barrington Moore, *Social Origins of Dictatorship and Democracy: Lord and Peasant in the Making of the Modern World* (Boston: Beacon Press, 1966), pp. 162–71; and Andrew Wilson, "Water, Power, and Culture in the Roman and Byzantine Worlds," *Water History*, vol. 4, no. 1 (April 2012), pp. 1–9.

53. Part of this shift was also driven by the sheer magnitude of population growth. For tens of thousands of years, the total number of humans on earth had fluctuated between an estimated 5 and 8 million. By around 4000 B.C., however, and after several millennia of farming, the world's population had risen to between 60 and 70 million. See *World Civilizations: The Origins of Civilizations. A project by History World International*, at http://history-world.org/neolithic1.htm.

54. Frances Gies and Joseph Gies, *Marriage and the Family in the Middle Ages* (New York: Harper & Row, 1987), p. 1.

55. See Sue Blundell, *Women in Ancient Greece* (Cambridge, MA: Harvard University Press, 1995), pp. 114–26.

56. See Frederick Engels, *The Origin of the Family, Private Property and the State*, p. 57; and Blundell, *Women in Ancient Greece*, p. 139.

57. Engels, *The Origin of the Family, Private Property and the State*, p. 23.

58. For an interesting discussion of the extent to which a man's honor throughout this period was connected to his female relatives' virginity, see Graeber, *Debt*, pp. 176–86.

59. David Herlihy, *Medieval Culture and Society* (New York: Harper & Row, 1968), p. 2.

60. For more on the economics behind this family-based structure, see Robert Fossier, "The Feudal Era (Eleventh–Thirteenth Century)," in André Burguière et al., eds., *A History of the Family* (Cambridge, UK: Polity Press, 1996), pp. 407–29.

61. There were, of course, exceptions to these rules, and intrigues that ranged across time and place. For an overview of marriage during the Middle Ages, see Coontz, *Marriage, a History*, pp. 88–103.

62. Gies and Gies, *Marriage and the Family in the Middle Ages*, p. 87.

63. I am not arguing here, as others have, that the agricultural surpluses of the Neolithic Revolution allowed religion itself to emerge, but rather that these surpluses and the societies in which they grew were subsequently embedded in and linked to particular religious expression. For a broader discussion of these relationships, see Ian Hodder, ed., *Religion at Work in a Neolithic Society* (Cambridge, UK: Cambridge University Press, 2014), esp. pp. 1–32.

64. The schism of 1054 split this empire into two, with the Roman pope presiding over western Europe's increasingly Christian population, and the ecumenical patriarch presiding over the continent's Eastern Orthodox followers.

65. M. Farah, *Marriage and Sexuality in Islam: A Translation of al-Ghazali's Book on the Etiquette of Marriage from the Ihya* (Salt Lake City: University of Utah Press, 1984), p. 26.

66. For example, Philippians 4:6–7.

67. Mark 10:9.

68. I Cor. 7:5. Nevertheless, Coontz notes that out-of-wedlock births were actually quite common among the peasantry of western Europe. See Coontz, *Marriage, a History*, p. 112.

69. See Sara McDougall, "The Transformation of Adultery in France at the End of the Middle Ages," *Law and History Review*, vol. 32, no. 3 (August 2014), pp. 491–524.

70. This custom took a specific turn in England, where, under the fourteenth-century laws of primogeniture, land could be granted only to one child—usually the eldest son. Historians have argued that the consolidation of land, and wealth, that accrued as a result helped Britain's subsequent rise to power. For a description of primogeniture and its effects, see Lawrence Stone, *The Family, Sex and Marriage in England, 1500–1800* (New York: Harper & Row, 1977), pp. 87–91.

71. Some historians have speculated that the slow pace of technological change during this era also enabled the ruling classes of Europe and Asia to keep the peasants on their land: without any hope of using technology to drive productivity, the argument goes, kings and queens were simply better off by taxing the peasants and hoping that small increases in population would, over time, increase the labor working the land and thus, in turn, their wealth. See Pierre Toubert, "The Carolingian Moment (Eighth–Tenth Century)," in André Burguière et al., eds., *A History of the Family*, pp. 379–406.

72. See White, *Medieval Technology and Social Change*, for an excellent overview. White essentially argues that technology evolved slowly throughout the Middle Ages, with sequential advances coming into more obvious fruition around the fourteenth century.

2. Steam Heat

1. Eric Hobsbawm gives a beautiful synopsis of the scope of movement prior to the revolution. See Eric Hobsbawm, *The Age of Revolution 1789–1848* (New York: Vintage Books, 1996), pp. 9–10.
2. Karl Marx, *The Poverty of Philosophy* (New York: International Publishers, 1963), p. 109.
3. The literature on the Industrial Revolution is voluminous. Some of the best sources, and the ones that provide the backbone for this chapter, include David S. Landes, *The Unbound Prometheus: Technological Change and Industrial Development in Western Europe from 1750 to the Present* (Cambridge, UK: Cambridge University Press, 1969); Peter N. Stearns, *The Industrial Revolution in World History* (Boulder, CO: Westview Press, 1998); and T. S. Ashton, *The Industrial Revolution, 1760–1830* (Oxford: Oxford University Press, 1997).
4. See, for instance, Adam Robert Lucas, "Industrial Milling in the Ancient and Medieval Worlds: A Survey of the Evidence for an Industrial Revolution in Medieval Europe," *Technology and Culture*, vol. 46, no. 1 (January 2005), pp. 1–30; and Joel Mokyr, *The Lever of Riches: Technological Creativity and Economic Progress* (New York: Oxford University Press, 1990), esp. pp. 31–56.
5. See Peter Mathias, "Financing the Industrial Revolution," in Peter Mathias and John A. Davis, eds., *The First Industrial Revolutions* (Cambridge, MA: Basil Blackwell, 1990), pp. 69–85; Anne L. Murphy, "The Financial Revolution and Its Consequences," in Roderick Floud, Jane Humphries, and Paul Johnson, eds., *The Cambridge Economic History of Modern Britain*, vol. 1 (Cambridge, UK: Cambridge University Press, 2014), pp. 321–43; and Hobsbawm, *The Age of Revolution 1789–1848*, esp. pp. 18–22.
6. Agricultural change during this period was not limited to the potato, and included both a broader use of fodder crops and evolving techniques of crop rotation. See Mark Overton, "Land and Labour Productivity in English Agriculture, 1650–1850," in Peter Mathias and John A. Davis, *The Nature of Industrialization*: vol. 4, *Agriculture and Industrialization* (Oxford: Blackwell Publishers, 1996), pp. 17–39.
7. See William H. McNeill, "How the Potato Changed the World's History," *Social Research*, vol. 66, no. 1 (Spring 1999), pp. 67–83; and Stearns, *The Industrial Revolution in World History*, p. 19. For a discussion of the horrors that ensued in Ireland when potatoes did fall prey to a devastating blight, see Cecil Woodham-Smith, *The Great Hunger: Ireland 1845–1849* (London: Penguin Books, 1991); and John Percival, *The Great Famine: Ireland's Potato Famine, 1845–1851* (New York: Viewer Books, 1995).
8. Stearns, *The Industrial Revolution in World History*, p. 19.
9. See, for example, Karl Marx, "Impact of Agricultural Revolution on Industry. The Creation of a Home Market for Industrial Capital," *Capital: A Critique of Political Economy*, vol. 1, trans. Ben Fowkes (London: Penguin, in association with New Left Review, 1990), pp. 908–13.
10. For a masterful account of cotton's history and impact, see Sven Beckert, *Empire*

of Cotton: A Global History (New York: Vintage, 2014). The next few paragraphs draw heavily on this work.

11. From *Cotton Supply Reporter*, no. 37 (March 1, 1860), p. 33, quoted in Beckert, *Empire of Cotton*, p. xviii.

12. There were several variations of this trade, but in general European merchants purchased slaves in Africa; shipped them to the American colonies or West Indies and sold them for crops such as cotton, sugar, or rum; and then imported these commodities back to Europe. The proceeds from their European sales could then be used to furnish funds for additional slave purchases.

13. About 12 million slaves entered the Atlantic trade between the sixteenth and nineteenth centuries, and an estimated 1.5 million died on board the slave ships. See Patrick Manning, "The Slave Trade: The Formal Demographics of a Global System," in Joseph E. Inikori and Stanley L. Engerman, eds., *The Atlantic Slave Trade: Effects on Economies, Societies and Peoples in Africa, the Americas, and Europe* (Durham, NC: Duke University Press, 1992), pp. 117–44.

14. Some of this demand had already been primed by imports of cloth from India, which also began to increase over the course of the eighteenth century. See Beckert, *Empire of Cotton*, p. 35.

15. Marxist historians have long viewed enclosure, and especially the practices that prevailed around the late eighteenth century, as one of the central causes (and calamities) of the Industrial Revolution. See E. P. Thompson, *The Making of the English Working Class* (New York: Vintage Books, 1966), esp. pp. 198–99.

16. Beckert, *Empire of Cotton*, p. 57.

17. David Landes, one of the most respected historians of the Industrial Revolution, argues that the division of labor that occurred around Manchester, and in the textile industry more generally, was one of the defining and enabling characteristics of early capitalism. See David S. Landes, "Introduction," in David Landes, ed., *The Rise of Capitalism* (New York: Macmillan, 1966), pp. 1–25.

18. Beckert, *Empire of Cotton*, p. 73. See also Landes, *The Unbound Prometheus*, pp. 84–88.

19. Beckert, *Empire of Cotton*, p. 63.

20. For Watt's life and history, along with stories of other significant pioneers in the creation of steam power, see Maury Klein, *The Power Makers: Steam, Electricity, and the Men Who Invented Modern America* (New York: Bloomsbury Press, 2008), esp. pp. 14–29.

21. See John U. Nef, "An Early Energy Crisis and Its Consequences," *Scientific American*, November 1977, pp. 140–51.

22. One early inventor was Edward Somerset, second Marquis of Worcester, who published a book with several ideas for steam-powered fountains. See Edward Somerset, Marquis of Worcester, *The Century of Inventions of the Marquis of Worcester. From the Original Ms. With Historical and Explanatory Notes and a Biographical Memoir by Charles F. Partington* (London: J. Murray, 1825). A more complex theory of the mechanics was laid out in a 1690 paper by Denis Papin, titled "Nouvelle méthode pour obtenir à bas prix des forces considérables" (A new method for cheaply obtaining considerable forces). See Louis Figuer, *Merveilles de la Science* (Paris: Jouvet et Cie, 1867), pp. 42–63.

23. For a description of the specific mechanisms and their development, see Harry Kitskopoulos, "From Hero to Newcomen: The Critical Scientific and Technological Developments That Led to the Invention of the Steam Engine," *Proceedings of the American Philosophical Society*, vol. 157, no. 3 (2013), pp. 304–44, at https://www.jstor.org/stable/24640398, accessed November 2019.

24. From Klein, *The Power Makers*, p. 25.

25. Ibid.; and Richard L. Hills, *Power from Steam: A History of the Stationary Steam Engine* (Cambridge, UK: Cambridge University Press, 1989), pp. 75, 87–88.

26. Beckert, *Empire of Cotton*, p. 65.

27. Ibid.

28. Hills, *Power from Steam*, p. 62; and Klein, *The Power Makers*, p. 26.

29. Klein, *The Power Makers*, p. 26.

30. Ibid.

31. The employment figures include only the major Western producing countries and do not include slaves. From Beckert, *Empire of Cotton*, pp. 180, 205–206.

32. Ibid., p. 68.

33. Landes, *The Unbound Prometheus*, p. 120.

34. Ibid., p. 2.

35. Much has been written about the lives of those who worked in the mills, and the changing nature of how time and work were measured there. See, for instance, Stearns, *The Industrial Revolution in World History*, pp. 57–61; E. P. Thompson, "Time, Work-Discipline, and Industrial Capitalism," *Past and Present*, no. 38 (December 1967), pp. 56–97; and Ivy Pinchbeck, *Women Workers and the Industrial Revolution: 1750–1850* (New York: Augustus M. Kelley Publishers, 1969), pp. 183–201. For firsthand accounts, see Simon Smike, *The Factory Lad: Or, the Life of Simon Smike, Exemplifying the Horrors of White Slavery* (London: T. White, 1839); and Mary Merryweather, *Experience of Factory Life*, 3rd ed. (London: Victoria Press, 1862).

36. For background on industrialization in Europe, see generally W. O. Henderson, *The Industrial Revolution in Europe: Germany, France, Russia, 1815–1914* (Chicago: Quadrangle Books, 1961).

37. Karl Marx, "Speech at the Anniversary of the People's Paper," 1856, in Robert C. Tucker, ed., *The Marx-Engels Reader*, 2nd ed. (New York: W. W. Norton, 1978), pp. 577–78.

38. See Friedrich Engels, "The Origin of the Family, Private Property, and the State," in Tucker, *The Marx-Engels Reader*, p. 739; and Karl Marx, "The German Ideology," in Tucker, *The Marx-Engels Reader*, pp. 146–200.

39. Engels, "The Origin of the Family, Private Property, and the State," p. 744.

40. Ibid., p. 742.

41. See, for example, Lawrence Stone, *The Family, Sex and Marriage in England, 1500–1800* (New York: Harper & Row, 1977), pp. 199–200.

42. For illustrative descriptions, see Laura Levine Frader, "Women in the Industrial Capitalist Economy," in Renate Bridenthal, Claudia Koonz, and Susan Stuard, eds., *Becoming Visible: Women in European History* (Boston: Houghton Mifflin, 1987), pp. 309–33; Pinchbeck, *Women Workers and the Industrial Revolution: 1750–1850*; and

more generally John Rule, *The Labouring Classes in Early Industrial England 1750–1850* (London: Longman, 1986).

43. Harriet Farley, quoted in Harriet H. Robinson, *Loom and Spindle: Or Life Among the Early Mill Girls* (Kailua, HI: Press Pacifica, 1976), p. 87. Even Thompson, a fierce critic of the mills and their effect, notes that "the abundant opportunities for female employment in the textile districts gave to women the status of independent wage earners." See Thompson, *The Making of the English Working Class*, pp. 303, 414.

44. See Joan W. Scott and Louise A. Tilly, "Women's Work and the Family in Nineteenth-Century Europe," *Comparative Studies in Society and History*, vol. 17, no. 1 (January 1975), pp. 36–64; Rule, *The Labouring Classes in Early Industrial England 1750–1850*, pp. 177–78; and Stone, *The Family, Sex and Marriage in England, 1500–1800*, p. 662.

45. For changes in the labor force over this period, see Pinchbeck, *Women Workers and the Industrial Revolution: 1750–1850*, pp. 116–17.

46. Ibid.

47. Peter Mathias, "Agriculture and Industrialization," in Mathias and Davis, eds., *The First Industrial Revolutions*, pp. 101–26.

48. *Report and Resolutions of a Meeting of Deputies from the Hand-Loom Worsted Weavers residing in and near Bradford, Leeds, Halifax, &c.* (1835), cited in Thompson, *The Making of the English Working Class*, p. 303.

49. Quoted in Pinchbeck, *Women Workers and the Industrial Revolution: 1750–1850*, p. 200.

50. Letter from the Printers' Union to the editor of *Philadelphia Daily News*, 1854, reproduced in Rosalyn Baxandall and Linda Gordon, eds., *America's Working Women* (New York: W. W. Norton, 1995), pp. 77–78. See also Joan W. Scott and Louise A. Tilly, "Women's Work and the Family in Nineteenth-Century Europe," *Comparative Studies in Society and History*, vol. 17, no. 1 (January 1975), pp. 36–64; and Rule, *The Labouring Classes in Early Industrial England 1750–1850*, pp. 177–78.

51. Muller v. Oregon, 208 U.S. 412, at 422. See also Ava Baron, "Protective Labor Legislation and the Cult of Domesticity," *Journal of Family Issues*, vol. 2, no. 1 (March 1981), pp. 25–38.

52. Rule, *The Labouring Classes in Early Industrial England 1750–1850*, pp. 178–79.

53. Nigel Goose, *Women's Work in Industrial England: Regional and Local Perspectives* (Hatfield, Hertfordshire: Local Population Studies, 2007), Table 7.2, pp. 168–69; and Stone, *The Family, Sex and Marriage in England, 1500–1800*, p. 201.

54. For a fascinating account of this family structure, which Lawrence Stone labels and describes as the "open lineage family," see Stone, *The Family, Sex and Marriage in England, 1500–1800*, pp. 85–119.

55. According to Stone, the shift toward what he calls the "closed domesticated nuclear family" emerged from the related growth of "affective individualism"—a new interest in the self, rather than the wider clan or kin grouping. Although Stone attributes this change to a wide array of cultural, political, economic, and ideological factors, one of the most important was the extent to which science in

the seventeenth century began to replace religion as a central organizing principle, and to suggest that "passive acceptance of one's fate was no longer the only . . . response to problems of life on earth." Stone, *The Family, Sex and Marriage in England, 1500–1800*, p. 233.

56. For a wrenching analysis of this period, see Thompson, *The Making of the English Working Class*.

57. There is substantial controversy among economic historians about just how rapid this growth was, and how it was divided among various portions of the population. The most optimistic estimates are associated with John Clapham and T. S. Ashton; the more pessimistic, with Marx, Engels, and their followers. For overviews and opinions, see E. J. Hobsbawm, "The British Standard of Living 1790–1850," *Economic History Review*, New Series, vol. 10, no. 1 (1957); A. J. Taylor, "Progress and Poverty in Britain 1780–1850," *History*, XLV (1960); and Maxine Berg, "Revisions and Revolutions: Technology and Productivity Change in Manufacture in Eighteenth-Century England," in Peter Mathis and John A. Davis, eds., *Innovation and Technology in Europe* (Oxford: Basil Blackwell, 1991), pp. 43–64. For an argument that the working classes suffered greatly despite the uptick in their material standards, see Thompson, *The Making of the English Working Class*, esp. pp. 189–212.

58. Hobsbawm, "The British Standard of Living 1790–1850," p. 47. For a careful review of the early data, see also Phyllis Deane, "Contemporary Estimates of National Income in the Second Half of the Nineteenth Century," *Economic History Review*, New Series, vol. 9, no. 3 (1957), pp. 451–61.

59. Interestingly, extended-family living remained more common among the aristocracy, including the royal family.

60. According to Stone and other historians of this period, the norm of doting on one's children rather than "ruthlessly crushing . . . their wills" was also a recent development, and one that paralleled the embrace of the nuclear family. See Stone, *The Family, Sex and Marriage in England, 1500–1800*, pp. 405–80.

61. Barbara Corrado Pope, "Angels in the Devil's Workshop: Leisured and Charitable Women in Nineteenth Century England and France," in Bridenthal, Koonz, and Stuard, eds., *Becoming Visible: Women in European History*, p. 303. Also see Hilary Land, "The Family Wage," *Feminist Review*, no. 6 (1980), pp. 55–77.

62. See Deborah Simonton, *A History of European Women's Work, 1700 to the Present* (London: Routledge, 1998), pp. 88–90. Or as Lawrence Stone writes (somewhat dismissively): "Among the upper tradesmen and shopkeepers, yeomen and tenant farmers, more and more wives were being educated in the social graces of their betters, and were withdrawing from active participation in family economic production." Stone, *The Family, Sex and Marriage in England, 1500–1800*, p. 656.

63. Virginia Woolf, "Professions for Women," in Mitchel A. Leaska, ed., *The Virginia Woolf Reader* (New York: Harcourt Brace Jovanovich, 1984), p. 278.

64. See Barbara Welter, "The Cult of True Womanhood: 1820–1860," *American Quarterly*, vol. 18, no. 2 (Summer 1966), pp. 151–74.

65. See, for example, Lorna Duffin, "The Prisoners of Progress: Women and Evolution," in Sarah Delamont and Lorna Duffin, eds., *The Nineteenth Century Woman:*

Her Cultural and Physical World (New York: Barnes & Noble Books, 1978), pp. 57–91; and Nancy Reagin, "The Imagined Hausfrau: National Identity, Domesticity, and Colonialism in Imperial Germany," *Journal of Modern History*, vol. 73, no. 1 (2001), pp. 54–86.

66. See Simonton, *A History of European Women's Work, 1700 to the Present*, p. 88. For additional description and analysis, see Stone, *The Family, Sex and Marriage in England, 1500–1800*, pp. 358–60.

67. John Ruskin, *Sesame and Lilies* (Chicago: Scott, Foresman, 1906), p. 137.

68. From *Cassell's Household Guide to Every Department of Practical Life: Being a Complete Encyclopaedia of Domestic and Social Economy*, vol. 1 (London: Cassell, Petter, Galpin, 1869), p. 1.

69. Eric Richards, "Women in the British Economy Since About 1700: An Interpretation," *History*, vol. 59, no. 197 (1974), p. 349.

70. The figure was reported by Mrs. Beeton's husband, who was also her publisher, so one might want to view it with some skepticism. See Kathryn Hughes, *The Short Life and Long Times of Mrs. Beeton* (New York: Alfred A. Knopf, 2006), p. 6.

71. Isabella Beeton, *The Book of Household Management* (New York: Farrar, Straus and Giroux, 1977), p. 1.

72. See *Guide des femmes de ménage, des cuisinières, et des bonnes d'enfants*, published by Simon-François Blocquel and available at http://gallica.bnf.fr/ark:/12148/bpt6k65645995; and *Schweizer Frauenheim*, available at https://www.swissbib.ch/Record/260830186/Description#tabnav.

73. Simonton, *A History of European Women's Work, 1700 to the Present*, p. 92.

74. Hills, *Power from Steam*, p. 117.

75. See Douglass North's 1993 Nobel Prize lecture, at http://www.nobelprize.org/nobel_prizes/economic-sciences/laureates/1993/north-lecture.html.

76. These questions captivated learned circles during the industrial era and have boggled historians ever since. See, for example, the summary in E. J. Hobsbawm, "The Standard of Living During the Industrial Revolution: A Discussion," *Economic History Review*, New Series, vol. 16, no. 1 (1963), pp. 119–34.

77. Marx, *Capital*, vol. 1, p. 621.

78. See Jeffrey Forgeng, *Daily Life in Stuart England* (Westport: Greenwood Press, 2007), p. 119.

3. Mid-Century Modern

1. Quoted in Ruth Schwartz Cowan, *More Work for Mother: The Ironies of Household Technologies from the Open Hearth to the Microwave* (New York: Basic Books, 1983), p. 84.

2. For descriptions of some earlier, and ill-fated, creations, see Herbert Lee Barber, *The Story of the Automobile: Its History and Development from 1760 to 1917* (Chicago: A. J. Munson, 1917), pp. 49–76.

3. The literature on Ford and his Model T is voluminous. For two very different studies, see Richard S. Tedlow, *New and Improved: The Story of Mass Marketing in America* (New York: Basic Books, 1990); and Barber, *The Story of the Automobile*.

4. Quoted in Tedlow, *New and Improved*, p. 120.
5. A price of $950 is roughly twenty-five thousand dollars in 2019 dollars.
6. By contemporary standards, this was still a complicated and somewhat arduous endeavor. But at the time, it represented a radical breakthrough in convenience.
7. Barber, *The Story of the Automobile*, p. 100.
8. See "Who Made America? Henry Ford," PBS, at https://pbs.org/wgbh/theymade america/whomade/ford_hi.html.
9. Ibid.; and Tedlow, *New and Improved*, p. 125.
10. Montgomery Rollins, quoted in Michael L. Berger, "Women Drivers! The Emergence of Folklore and Stereotypic Opinions Concerning Feminine Automotive Behavior," *Women's Studies International Forum*, vol. 9, no. 3 (1986), p. 260.
11. Montgomery Rollins, quoted in ibid., p. 261.
12. See "Farm Women Who Count Themselves Blest by Fate," *Literary Digest*, November 13, 1920, p. 52.
13. Deborah Clarke, *Driving Women: Fiction and Automobile Culture in Twentieth-Century America* (Baltimore: Johns Hopkins University Press, 2007), p. 10.
14. Mrs. A. Sherman Hitchcock, "A Woman's Viewpoint of Motoring," *Motor*, April 1904, p. 19.
15. Mrs. Andrew Cuneo, quoted in Robert Sloss, "What a Woman Can Do with an Auto," *Outing*, April 1910, p. 69.
16. Clarke, *Driving Women*, p. 15.
17. See Peter J. Ling, *America and the Automobile: Technology, Reform and Social Change* (Manchester: Manchester University Press, 1990), pp. 37–94. Earlier suburbs had similarly developed in the mid-nineteenth century, clustered around railroad and streetcar routes. See Sam Bass Warner, *Streetcar Suburbs* (Cambridge, MA: Harvard University Press, 1962); and Robert J. Jucha, "The Anatomy of a Streetcar Suburb," *Western Pennsylvania History Magazine*, vol. 62, no. 4 (October 1979), pp. 301–19.
18. Margaret Walsh, "Gender and the Automobile in the United States: Consumerism and the Great Economic Boom," *The Automobile in American Life and Society* (Dearborn: Henry Ford Museum and University of Michigan, 2005), at http://www.autolife.umd.umich.edu/Gender/Walsh/G_Overview3.htm.
19. See *Survey of Current Business*, December 1961, Table of General Business Indicators, p. S-1, at https://bea.gov/scb/pdf/1961/1261cont.pdf; and Walsh, "Gender and the Automobile in the United States." These patterns of ownership were by no means distributed evenly or equitably across the country. African Americans, in particular, faced far greater difficulties in buying and financing their vehicles, and had, as a result, considerably lower levels of ownership. For a description and analysis, see Thomas J. Sugrue, "Driving While Black: The Car and Race Relations in Modern America," in *The Automobile in American Life and Society*, at http://www.autolife.umd.umich.edu/Race/R_Casestudy/R_Casestudy.htm; and Warren Brown, "Cadillac's Cultural Turn," *Washington Post*, December 24, 1995, at https://www.washingtonpost.com/archive/business/1995/12/24/cadillacs-cultural-turn/7374f4c7-b78f-4007-9938-51ab24bf3522/?noredirect=on&utm_term=.45752f54bcf9.

20. Women were 40 percent of licensed drivers in 1963. See U.S. Department of Transportation Federal Highway Administration, *Highway Statistics Summary to 1995*, April 1997, Table DL-220, at https://www.fhwa.dot.gov/ohim/summary95/dl220.pdf.
21. See Walsh, "Gender and the Automobile in the United States."
22. For a beautiful account written at the time, see Robert S. Lynd and Helen Merrell Lynd, *Middletown: A Study in American Culture* (New York: Harcourt, Brace & World, 1929), pp. 137–52.
23. Some historians argue that there was actually more sex occurring during the Victorian era than is commonly thought—and certainly more than commentators during the period were willing to acknowledge. See Carl N. Degler, "What Ought to Be and What Was: Women's Sexuality in the Nineteenth Century," *American Historical Review*, vol. 79, no. 5 (December 1974), pp. 1467–90; and Caroll Smith-Rosenberg, "Sex as Symbol in Victorian Purity: An Ethnohistorical Analysis of Jacksonian America," *American Journal of Sociology*, vol. 84, Supplement: Turning Points: Historical Sociological Essays on the Family (1978), pp. S212–47.
24. See, for example, Cas Wouters, "No Sex Under My Roof: Teenage Sexuality in the USA and the Netherlands since the 1880s," *Política y Sociedad*, vol. 50, no. 2 (2013), pp. 421–52.
25. See, for example, Rufus S. Lusk, "The Drinking Habit," *Annals of the American Academy of Political and Social Science*, vol. 163, no. 1 (September 1932), pp. 46–52; and F. Eugene Melder, "The 'Tin Lizzie's' Golden Anniversary," *American Quarterly*, vol. 12, no. 4 (Winter 1960), pp. 466–81.
26. Winston Ehrmann, "Changing Sexual Mores," compiled by Eli Ginzburg in *Values and Ideals of American Youth* (New York: Columbia University Press, 1961), 56.
27. Walsh, "Gender and the Automobile in the United States."
28. Jonathan Rees, *Refrigeration Nation* (Baltimore: Johns Hopkins University Press, 2013), p. 138.
29. Ibid., p. 142.
30. The story of electricity and its development goes beyond the scope of this narrative. For excellent accounts of this long and colorful history, see Maury Klein, *The Power Makers: Steam, Electricity, and the Men Who Invented Modern America* (New York: Bloomsbury Press, 2008); Herbert W. Meyer, *A History of Electricity and Magnetism* (Cambridge, MA: MIT Press, 1971); L. Pearce Williams, *Michael Faraday: A Biography* (London: Chapman and Hall, 1965); and Thomas P. Hughes, *Networks of Power: Electrification in Western Society, 1880–1930* (Baltimore: Johns Hopkins University Press, 1983).
31. Dryers came along somewhat later, but had less of an impact, both in terms of sales penetration and workload reduction. For decades, many households continued to dry their laundry on lines set outside the home.
32. For contemporary descriptions of this process and how to manage it, see Catherine Esther Beecher, *A Treatise on Domestic Economy: for the Use of Young Ladies at Home, and at School* (New York: Harper, 1848); and L. Ray Balderston and M. C. Limerick, *Laundry Manual* (Philadelphia: Anvil Printing, 1902).

33. See Cowan, *More Work for Mother*, pp. 69–102.

34. Ibid., p. 98.

35. Ibid., p. 94; and J. Bradford DeLong, "The Roaring Twenties," in *Slouching Towards Utopia? The Economic History of the Twentieth Century* (February 1997), at http://holtz.org/Library/Social%20Science/Economics/Slouching%20Towards%20Utopia%20by%20DeLong/Slouch_roaring13.html.

36. DeLong, "The Roaring Twenties."

37. Betty Friedan, *The Feminine Mystique* (New York: Penguin Books, 2010), esp. pp. 167, 185–86; Naomi Wolf, *The Beauty Myth* (New York: Harper Perennial, 2002), pp. 64–65; and Dolores Hayden, *The Grand Domestic Revolution: A History of Feminist Designs for American Homes, Neighborhoods, and Cities* (Cambridge, MA: MIT Press, 1981). See also the discussion in David E. Nye, *Electrifying America: Social Meanings of a New Technology, 1880–1940* (Cambridge, MA: MIT Press, 1990), pp. 238–86.

38. Julie Wosk, *Women and the Machine: Representations from the Spinning Wheel to the Electronic Age* (Baltimore: Johns Hopkins University Press, 2001), p. 227.

39. Ibid., p. 229.

40. F. Molnar, "The New Rhythm of the American Economy (Cyclicality and Intermittence in Economic Development 1947–1971)," *Acta Oeconomica*, vol. 12, no. 2 (1974), pp. 206–207.

41. *Ladies' Home Journal*, vol. 66, no. 12 (December 1949), p. 32; and the *Saturday Evening Post*, vol. 232, no. 38 (March 19, 1960), p. 20.

42. Friedan, *The Feminine Mystique*, p. 277.

43. Much (but not all) of this increase came from women returning to the workforce once their children were older. See Elizabeth K. Waldman, "Changes in the Labor Force Activity of Women," *Monthly Labor Review*, vol. 93, no. 6 (June 1970), pp. 10–18.

44. Cowan, *More Work for Mother*, pp. 102–50.

45. From *Statistical Abstract of the United States, 1971*, Tables Number 328 and 329, at https://www.census.gov/library/publications/1971/compendia/statab/92ed.html.

46. Data from the OECD, at https://data.oecd.org/emp/labour-force.htm#indicator-chart.

47. The first explicit Christian attack on birth control came in A.D. 418, when St. Augustine wrote that "there is no matrimony where motherhood is prevented; for then there is no wife." For the history of legal prohibitions in the United States, see Andrea Tone, *Devices and Desires: A History of Contraceptives in America* (New York: Hill and Wang, 2001).

48. This section draws generally from Debora L. Spar and Briana Huntsberger, "Midwives, Witches, and Quacks: The Business of Birth Control in the Pre-Pill Era," Harvard Business School, working paper no. 04-049, 2004. For specific data, see Richard A. Easterlin, "The Economics and Sociology of Fertility: A Synthesis," in Charles Tilly, ed., *Historical Studies of Changing Fertility* (Princeton: Princeton University Press, 1978), pp. 69–79; Barbara Hanawalt, *The Ties That Bound: Peasant Families in Medieval England* (New York: Oxford University

Press, 1986), pp. 101–103; and John Riddle, *Eve's Herbs: A History of Contraception and Abortion in the West* (Cambridge, MA: Harvard University Press, 1997), pp. 18–20.

49. Spar and Huntsberger, "Midwives, Witches, and Quacks," pp. 14–15; and Charles Knowlton, *Fruits of Philosophy* (London: F. P. Rogers, Printer, 1839), pp. 33–36.

50. Spar and Huntsberger, "Midwives, Witches, and Quacks," p. 16; and John M. Riddle, *Contraception and Abortion from the Ancient World to the Renaissance* (Cambridge, MA: Harvard University Press, 1992), p. 5.

51. The central figure in this story and eventually in the history of the pill was a devout Catholic researcher named John Rock. For more on his life and work, see Loretta McLaughlin, *The Pill, John Rock, and the Church: The Biography of a Revolution* (Boston: Little, Brown, 1982); and Malcolm Gladwell, "John Rock's Error: What the Co-Inventor of the Pill Didn't Know," *New Yorker*, March 13, 2000. Also, and more generally, see Debora L. Spar, *The Baby Business: How Money, Science, and Politics Drive the Commerce of Conception* (Boston: Harvard Business School Press, 2006).

52. Spar and Huntsberger, "Midwives, Witches, and Quacks," pp. 33–38.

53. See Dagmar Herzog, *Sexuality in Europe: A Twentieth Century History* (Cambridge, UK: Cambridge University Press, 2011), esp. p. 139.

54. Claudia Goldin and Lawrence F. Katz, "The Power of the Pill: Oral Contraceptives and Women's Career and Marriage Decisions," *Journal of Political Economy*, vol. 110, no. 4 (2002), p. 747.

55. Ibid., pp. 730–70. In a separate paper, Goldin and Katz argue that the pill created a social multiplier effect, by which marriage prospects are enhanced even for women who do not take the pill. See Claudia Goldin and Lawrence F. Katz, "Career and Marriage in the Age of the Pill," *American Economic Review*, vol. 90, no. 2 (May 2000), pp. 461–65.

56. In the United States, the percentage of female lawyers and judges rose from 5.1 percent in 1970 to 13.6 percent in 1980, while the share of female physicians increased from 9.1 percent in 1970 to 14.1 percent in 1980. See Goldin and Katz, "The Power of the Pill," p. 749. See also Martha J. Bailey, "More Power to the Pill: The Impact of Contraceptive Freedom on Women's Life Cycle Labor Supply," *Quarterly Journal of Economics*, vol. 121, no. 1 (February 2006), pp. 289–320; and Susan C. M. Scrimshaw, "Women and the Pill: From Panacea to Catalyst," *Family Planning Perspectives*, vol. 13, no. 6 (November–December 1981), pp. 254–56, 260–62.

57. The authors analogize this shift to the emergence of power looms, which similarly worsened prospects for those who did not adopt the new technology. George A. Akerlof, Janet L. Yellen, and Michael Katz, "An Analysis of Out-of-Wedlock Childbearing in the United States," *Quarterly Journal of Economics*, vol. 111, no. 2 (May 2, 1996), pp. 277–317. The quote is from page 309.

58. For a description of some of these arguments, see Andrew Hacker, "The Pill and Morality," *New York Times Magazine*, November 21, 1965, pp. 32, 138–40; Gloria Steinem, "The Moral Disarmament of Betty Coed," *Esquire*, September 1962, pp. 96–97, 153–57; and "The Pill: How It Is Affecting U.S. Morals, Family Life," *U.S. News & World Report*, July 11, 1966, pp. 62–65.

4. Changing the Means of (Re)production

1. "A Look at Global Feminism, with Two of Its Inventors," *New York Times*, November 3, 2017, p. C4.

2. "Resources," European Society of Human Reproduction and Embryology, at https://www.eshre.eu/Press-Room/Resources.aspx.

3. Ibid. Countries with the highest per capita use of assisted reproduction technologies include Belgium, Denmark, Finland, and Sweden.

4. The definitions here are complex and contentious. Technically, most data-collecting agencies define infertility as occurring when a heterosexual couple is unable to conceive a child after one year of unprotected sexual activity. Historically, demographers calculate its rate by the difference between marriage and birth rates: married couples who do not produce children are deemed to have been infertile. For further discussion of these definitions, see Elisabeth Hervey Stephen and Anjani Chandra, "Use of Fertility Services in the United States: 1995," *Family Planning Perspectives*, vol. 32, no. 3 (May–June 2000), pp. 132–37; and Maya N. Mascarenhas et al., "Measuring Infertility in Populations: Constructing a Standard Definition for Use with Demographic and Reproductive Health Surveys," *Population Health Metrics*, vol. 10 (August 2012), pp. 17–27.

5. See Debora L. Spar, *Wonder Women: Sex, Power, and the Quest for Perfection* (New York: Sarah Crichton Books, 2013), p. 135; and David Plotz, *The Genius Factory: The Curious History of the Nobel Prize Sperm Bank* (New York: Random House, 2006), pp. 159–62.

6. See Addison Davis Hard, "Artificial Impregnation," *The Medical World*, vol. 27 (April 1909), pp. 163–64; Elaine Tyler May, *Barren in the Promised Land: Childless Americans and the Pursuit of Happiness* (New York: Basic Books, 1995), p. 67; and Cynthia R. Daniels and Janet Golden, "Procreative Compounds: Popular Eugenics, Artificial Insemination, and the Rise of the American Sperm Banking Industry," *Journal of Social History*, vol. 38, no. 1 (Autumn 2004), p. 8.

7. It didn't end well. The bank's most well-known donor was reviled as a racist, and the only two other Nobel winners fled after the bank went public. For this, and other attempts at using selected sperm to create better babies, see Plotz, *The Genius Factory*; and Daniels and Golden, "Procreative Compounds," pp. 5–27.

8. Anne Taylor Fleming, "New Frontiers in Conception," *New York Times Magazine*, July 20, 1980, p. 14.

9. See Leon R. Kass, *Toward a More Natural Science: Biology and Human Affairs* (New York: Free Press, 1985), p. 114; and Paul Ramsay, *Fabricated Man: The Ethics of Genetic Control* (New Haven: Yale University Press, 1970), p. 138. Both these quotes and the substance of this paragraph are from Debora L. Spar, *The Baby Business: How Money, Science, and Politics Drive the Commerce of Conception* (Boston: Harvard Business School Press, 2006), p. 26.

10. Congregation for the Doctrine of the Faith, *Donum Vitae: Instructions on Respect for Human Life in Its Origins and the Dignity of Procreation* (London: Catholic Truth Society, 1987), p. 27. Quoted in Spar, *The Baby Business*, p. 26.

11. "Assisted Reproductive Technology in the United States and Canada: 1993 Results Generated from the American Society for Reproductive Medicine/Society for Assisted Reproductive Technology," *Fertility and Sterility*, vol. 64, no. 1 (July 1995), pp. 13–21; and Victoria Clay Wright et al., "Assisted Reproductive Technology Surveillance—United States, 2003," Division of Reproductive Health, National Center for Chronic Disease Prevention and Human Promotion, May 26, 2006/55(SS04), at https://www.cdc.gov/mmwr/preview/mmwrhtml/ss5504a1.htm.

12. Data from "6.5 Million IVF Babies Born Worldwide Since Louise Brown," The Infertility Journey, July 20, 2016, at https://www.theinfertilityjourney.com/post/2016/07/21/65-million-ivf-babies-born-worldwide-since-louise-brown.

13. Quoted in Liat Collins, "A Labor of Love," *Jerusalem Post*, September 14, 2001, p. 7; and cited in Spar, *The Baby Business*, p. 43.

14. Debora Spar, "Should You Freeze Your Eggs?" *Marie Claire*, July 2015, pp. 132–33, 146.

15. The most dramatic and well-known of these cases was the Baby M case, discussed later in this chapter. For more on the inevitable complications of gestational surrogacy arrangements, see J. Herbie DiFonzo and Ruth Stern, "The Children of Baby M," *Capital University Law Review*, vol. 39, no. 2 (Spring 2011), pp. 345–411; and Elizabeth Scott, "Surrogacy and the Politics of Commodification," *Law and Contemporary Problems*, vol. 72, no. 3 (Summer 2009), pp. 109–46.

16. A. H. Handyside et al., "Biopsy of Human Preimplantation Embryos and Sexing by DNA Amplification," *Lancet*, vol. 333, no. 8634 (February 18, 1989), pp. 347–49; and A. H. Handyside et al., "Pregnancies from Biopsied Human Preimplantation Embryos Sexed by Y-Specific DNA Amplification," *Nature*, vol. 344 (April 19, 1990), pp. 768–70.

17. For more on this episode, and a similar, but failed, attempt to use PGD to save another child's life, see Lisa Belkin, "The Made-to-Order Savior," *New York Times Magazine*, July 1, 2001, pp. 36–43; and Spar, *The Baby Business*, pp. 99–101.

18. Leon R. Kass, *Life, Liberty and the Defense of Dignity: The Challenge for Bioethics* (San Francisco: Encounter Books, 2002), p. 131.

19. Francis Fukuyama, *Our Posthuman Future: Consequences of the Biotechnology Revolution* (New York: Farrar, Straus and Giroux, 2002), p. 157.

20. Spar, *The Baby Business*, p. 115.

21. "Preimplantation Genetic Diagnosis: Experience of 3000 Clinical Cycles," Report of the 11th Annual Meeting of International Working Group on Preimplantation Genetics, May 15, 2001, *Reproductive Biomedicine Online*, vol. 3, no. 1 (2001), pp. 49–53, at https://www.rbmojournal.com/article/S1472-6483(10)61967-0/abstract.

22. The interviews were conducted by my research assistant Aryanna Garber.

23. Personal interview with researcher, August 9, 2017.

24. Personal interview with researcher, July 6, 2017. One year after the interview, the couple became dads to a healthy and beautiful son.

25. For a legal history, see Martha F. Davis, "Male Couverture: Law and the Illegiti-

mate Family," *Rutgers Law Review*, vol. 56, no. 1 (Fall 2003), pp. 73–118; and Serena Mayeri, "Marital Supremacy and the Constitution of the Nonmarital Family," *California Law Review*, vol. 103, no. 5 (October 2015), pp. 1277–352.

26. Ivy Pinchbeck, "Social Attitudes to the Problem of Illegitimacy," *British Journal of Sociology*, vol. 5, no. 4 (December 1954), pp. 309–23.

27. See Steve Olson, "Who's Your Daddy?" *Atlantic*, July–August 2007, at https://www .theatlantic.com/magazine/archive/2007/07/who-s-your-daddy/305969/; and Mark Pagel, *Wired for Culture: Origins of the Human Social Mind* (New York: W. W. Norton, 2012), p. 315.

28. "Illegitimacy," *International Encyclopedia of the Social Sciences*, at http://www .encyclopedia.com/social-sciences-and-law/law/law/illegitimacy.

29. *The Registrar General's Statistical Review of England and Wales, 1946–1950*, table XLIX (London: H.M.S.O., 1954), p. 89; and *The Registrar General's Statistical Review of England and Wales, 1968*, Part II, Tables, Populations (London: H.M.S.O., 1970).

30. See Jennifer Mittelstadt, *From Welfare to Workfare: The Unintended Consequences of Liberal Reform, 1945–1965* (Chapel Hill: University of North Carolina Press, 2005), pp. 86–91.

31. Memorandum from the ACLU on Louisiana Plan for Aid to Defendant Children, filed with the Department of Health, Education and Welfare (November 22, 1960), quoted in Mayeri, "Marital Supremacy and the Constitution of the Nonmarital Family," p. 1287.

32. Ibid.

33. Harry D. Krause, "Equal Protection for the Illegitimate," *Michigan Law Review*, vol. 65, no. 3 (January 1967), p. 488.

34. Ibid., p. 484.

35. Levy v. Louisiana, 391 U.S. 68 (1968), 70–72. See also John C. Gay and David Rudovsky, "The Court Acknowledges the Illegitimate: *Levy v. Louisiana* and *Glona v. American Guarantee & Liability Insurance Co.*," *University of Pennsylvania Law Review*, vol. 118, no. 1 (November 1969), pp. 1–39.

36. Glona v. American Guarantee and Liability Insurance Company, 391 U.S. 73 (1968).

37. Gomez v. Perez, 409 U.S. 535 (1973).

38. Stephanie Coontz, *Marriage, a History: From Obedience to Intimacy or How Love Conquered Marriage* (New York: Penguin Books, 2006), p. 257.

39. China explicitly protected illegitimate children under the Marriage Law of 1980. Japan granted illegitimate children partial access to their parents' inheritance in the 1980s, and then full rights in 2013.

40. In both cases, some critics have argued that focusing on the interests of the child de-emphasizes other, broader issues of equity, equality, and dignity. See Nancy D. Polikoff, "For the Sake of All Children: Opponents and Supporters of Same-Sex Marriage Both Miss the Mark," *The City University of New York Law Review*, vol. 8, no. 2 (Fall 2005), pp. 573–98.

41. *In re* Baby M, 537 A.2d 1227, 1234 (N.J. 1988). Since the decision, many critics have noted its biases. Apart from the complicated issues of surrogacy, the Sterns are repeatedly described as stable, wealthy, and professional—all deemed key aspects

of good parenting. See, for instance, "Baby M Verdict Neglected the Mother's Role," *New York Times*, Letter to the Editor, April 20, 1987, at http://www.nytimes .com/1987/04/20/opinion/I-baby-m-verdict-neglected-the-mother-s- role-908487 .html.

42. Johnson v. Calvert, 5 Cal. 4th 84, 851 P.2d 776 (Cal.1993); and Spar, *The Baby Business*, p. 85.

43. *Johnson*, 851 P.2d, at 782. For a broader analysis, see Douglas NeJaime, "Marriage Equality and the New Parenthood," *Harvard Law Review*, vol. 129, no. 5 (March 2016), pp. 1210–11.

44. *In re* Marriage of Buzzanca, 61 Cal.App.4th 1410, 72 Cal.Rptr. 2d 280 (Ct. App. 1998); and NeJaime, "Marriage Equality and the New Parenthood," p. 1211.

45. For full description of this period and its political fights, see Linda Harper, *Victory: The Triumphant Gay Revolution* (New York: Harper Perennial, 2012); David B. Feinberg, *Queer and Loathing: Rants and Raves of a Raging AIDS Clone* (New York: Viking, 1994); and George Chauncey, "The Long Road to Marriage Equality," *New York Times*, June 27, 2013, p. A31.

46. This period has been dubbed by some as the "lesbian baby boom." For a legal review of how lesbian advocates structured their fight in the California courts during this time, see NeJaime, "Marriage Equality and the New Parenthood," pp. 1196–208.

47. Seema Mohapatra, "Achieving Reproductive Justice in the International Surrogacy Market," *Annals of Health Laws*, vol. 21, no. 1 (2012), pp. 191–200.

48. Jack Glaser, "Womb for Rent: Regulating the International Surrogacy Market," *BPR Magazine*, November 6, 2016, at http://www.brownpoliticalreview.org/2016 /11/womb-for-rent-regulating-international-surrogacy-market/.

49. Media reports indicate that he paid around twenty-seven thousand dollars just for the surrogate carrier. See "Last of the Big Spenders! Elton John 'Paid £20,000' to Surrogate Mother for Giving Birth to Second Son Elijah," *Daily Mail*, January 20, 2013, at http://www.dailymail.co.uk/tvshowbiz/article-2265463/Elton -John-paid-20,000-surrogate-mother-giving-birth-second-son-Elijah.html.

50. California, where both Elton John and Neil Patrick Harris completed their surrogacy arrangements, has led the way in providing a more enabling legal environment. Since 2013, the California Family Code allows intended parents to apply for a pre-birth order that enables both parents—including same-sex parents—to be listed on the child's birth certificate. See CA Fam Code § 7962 (2013).

51. K.M. v. E.G., 117 P.3d 673 (Cal. 2005); and NeJaime, "Marriage Equality and the New Parenthood," pp. 1222–26.

52. Elisa B. v. Superior Court, 117 P.3d 660, 670 (Cal. 2005); and NeJaime, "Marriage Equality and the New Parenthood," pp. 1226–29.

53. Martin Kolk and Gunnar Andersson, "Two Decades of Same-Sex Marriage in Sweden: A Demographic Account," manuscript prepared for the 2016 meeting of the European Population Association, at https://epc2016.princeton.edu/papers /160627; and Elixabete Imaz, "Same-Sex Parenting, Assisted Reproduction and Gender Asymmetry: Reflecting on the Differential Effects of Legislation on Gay

and Lesbian Family Formation in Spain," *Reproductive BioMedicine & Society On-line*, vol. 4 (2017), pp. 5–12, at https://www.sciencedirect.com/science/article/pii/S2405661817300060.

54. For a full description and analysis of the connections between assisted reproduction technologies and the legal argument for marriage equality, see NeJaime, "Marriage Equality and the New Parenthood," pp. 1185–266.

55. Obergefell v. Hodges, 135 S. Ct. 2584, 2599, 2600 (2015).

56. *In re* Marriage Cases, 183 P.3d 384, 433 (Cal. 2008).

57. See Dan Bilefsky, "British Woman, 60, Wins Legal Round in Fight to Give Birth to Grandchild," *New York Times*, June 30, 2016, at https://www.nytimes.com/2016/07/01/world/europe/british-grandmother-60-wins-legal-round-in-fight-to-give-birth-to-grandchild.htm.

58. See Tom Payne, "Woman Who Gave Birth to Her Own BROTHER," *Daily Mail*, June 8, 2017, at http://www.dailymail.co.uk/news/article-4584222/Woman-gives-birth-BROTHER-surrogate.html.

59. See "City Freezes Eggs of Woman in Her 20s," *Japan Times*, June 18, 2016, p. 2.

60. Gina Hall, "These Tech Companies Pay for Egg-Freezing as an Employee Benefit," *Silicon Valley Business Journal*, May 23, 2017, at https://www.bizjournals.com/sanjose/news/2017/05/23/google-apple-facebook-intel-egg-freezing-benefit.html.

61. Recently, scientists have begun to hypothesize that eggs may be produced throughout a woman's life, rather than just being fixed from birth. See Joshua Johnson et al., "Germline Stem Cells and Follicular Renewal in the Postnatal Mammalian Ovary," *Nature*, vol. 428, no. 6979 (March 11, 2004), pp. 145–50.

62. For the key work in this area, see Kazutoshi Takahashi and Shinya Yamanaka, "Induction of Pluripotent Stem Cells from Mouse Embryonic and Adult Fibroblast Cultures by Defined Factors," *Cell*, vol. 126, no. 4 (August 25, 2006), pp. 663–76; In-Hyun Park et al., "Reprogramming of Human Somatic Cells to Pluripotency with Defined Factors," *Nature*, vol. 451 (January 2008), pp. 141–47; and Junying Yu et al., "Induced Pluripotent Stem Cell Lines Derived from Human Somatic Cells," *Science*, vol. 318, no. 5858 (December 21, 2007), pp. 1917–20.

63. Niels Geijsen et al., "Derivation of Embryonic Germ Cells and Male Gametes from Embryonic Stem Cells," *Nature*, vol. 427, no. 6970 (January 8, 2004), pp. 148–54; and Karin Hübner et al., "Derivation of Oocytes from Mouse Embryonic Stem Cells," *Science*, vol. 300, no. 5623 (May 23, 2003), pp. 1251–56.

64. Karim Nayernia et al., "In Vitro-Differentiated Embryonic Stem Cells Give Rise to Male Gametes That Can Generate Offspring Mice," *Developmental Cell*, vol. 11, no. 1 (July 2006), pp. 125–32; O. Hikabe et al., "Reconstitution *In Vitro* of the Entire Cycle of the Mouse Female Germ Line," *Nature*, vol. 539, no. 7628 (November 2016), pp. 299–303; and K. Hayashi et al., "Offspring from Oocytes Derived from *In Vitro* Primordial Germ Cell-like Cells in Mice," *Science*, vol. 338, no. 6109 (November 16, 2012), pp. 971–75. Also, and more generally, see A. J. Newsom and A. C. Smajdor, "Artificial Gametes: New Paths to Parenthood?" *Journal of Medical Ethics*, vol. 31, no. 3 (March 2005), pp. 184–86; César Palacios-González, John Harris, and Giuseppe Testa, "Multiplex Parenting: IVG and the Generations to Come,"

Journal of Medical Ethics, vol. 40, no. 11 (November 2014), pp. 752–58; and Hannah Bourne, Thomas Douglas, and Julian Savulescu, "Procreative Beneficence and *In Vitro* Gametogenesis," *Monash Bioethics Review*, vol. 30, no. 2 (September 2012), pp. 29–48.

65. As of 2017, the creation (and destruction) of IVG-derived human embryos solely for the purpose of research would be ineligible for public funding under the Dickey-Wicker Amendment, an appropriations rider that prohibits the use of federal funding for research wherein embryos are created for research purposes or destroyed. See I. Glenn Cohen, George Q. Daley, Eli Y. Adashi, "Disruptive Reproductive Technologies," *Science Translational Medicine*, vol. 9, no. 372 (January 11, 2017), at http://stm.sciencemag.org/content/9/372/eaag2959.

66. The exception, thus far, is cloning. Scientists have succeeded in cloning several large mammals—not only Dolly, the famous sheep, but also cows, pigs, and rabbits. At least in public, though, no one has claimed success in human cloning. For more on the links between animal and human cloning efforts, see Spar, *The Baby Business*, pp. 129–58.

67. Maggie Gallagher, in John Corvino and Maggie Gallagher, *Debating Same-Sex Marriage* (Oxford: Oxford University Press, 2012), p. 95.

68. With current technologies, there are a few ways of making this match. The simplest would be to use a woman's genetic material to create an egg cell that could be mated with a donor sperm. The resulting embryo could then generate its own sperm cell.

69. César Palacios-González, John Harris, and Giuseppe Testa, "Multiplex Parenting: IVG and the Generations to Come," *Journal of Medical Ethics*, vol. 40, issue 11 (November 2014), pp. 752–58.

70. Theoretically, a single woman could fertilize her own egg with sperm also produced from her own genetic material. Because this combination gets so close to cloning, however, it is not likely to become widely possible—or perhaps even permissible. For more on the possibilities of what the author terms "solo IVF," see Sonia M. Suter, "In Vitro Gametogenesis: Just Another Way to Have a Baby?" *Journal of Law and the Biosciences*, vol. 1, no. 3 (April 2016), pp. 106–10.

71. See Philip Lockley, ed., *Protestant Communalism in the Trans-Atlantic World, 1650–1850* (London: Palgrave Macmillan, 2016); Donald E. Pitzer, ed., *America's Communal Utopias* (Chapel Hill: University of North Carolina Press, 1997); and Clinton Nguyen, "6 Attempts at Utopian Settlements and Where Are They Now," *Business Insider*, October 25, 2016, at http://www.businessinsider.com/what-happened-to-utopian-settlements-2016-10/#the-harmony-society-existed-from-1805-to-1905-9.

72. See, for example, Palacios-González, Harris, and Testa, "Multiplex Parenting: IVG and the Generations to Come," pp. 752–58; and Suter, "In Vitro Gametogenesis: Just Another Way to Have a Baby?" pp. 106–10.

73. Already, a court in Australia has given a group of three—one man and two women—the legal standing of a civil union. See Ean Higgins, "Three in Marriage Bed More of a Good Thing," *Australian*, December 10, 2011, at http://www

.theaustralian.com.au/news/features/three-in-marriage-bed-more-of-a-good
-thing/story-e6frg6z6-1226218569577. In California, a 2013 law allows for more
than two legal parents, if not doing so would be detrimental to the child's interests. See Christopher Cadelago, "Jerry Brown Signs California Bill Allowing
More Than Two Parents," *Sacramento Bee*, October 4, 2013, at http://blogs.sacbee
.com/capitolalertlatest/2013/10/jerry-brown-signs-bill-allowing-more-than-two
-parents-in-calif.html.

5. Sex and Love Online

1. Cited in Nancy Jo Sales, "Tinder Is the Night," *Vanity Fair*, September 2015, available at http://www.vanityfair.com/culture/2015/08/tinder-hook-up-culture-end
   -of-dating. I have changed the wording slightly.
2. Data from Centers for Disease Control, reported in Tara Bahrampour, "There Isn't
   Really Anything Magical About It: Why More Millennials Are Avoiding Sex,"
   Washington Post, August 2, 2016, at https://www.washingtonpost.com/local/social
   -issues/there-isnt-really-anything-magical-about-it-why-more-millennials-are
   -putting-off-sex/2016/08/02/e7b73d6e-37f4-1136-87fc-d4c723a2becb_story.html;
   and Jean M. Twenge, Ryne A. Sherman, and Brooke E. Wells, "Sexual Inactivity
   During Young Adulthood Is More Common Among U.S. Millennials and iGen,"
   Archives of Sexual Behavior, vol. 46, no. 2 (February 2017), pp. 433–40. The latter data
   point comes from 2017.
3. See Kate Julian, "Why Are Young People Having So Little Sex?" *Atlantic*, December 2018, at https://www.theatlantic.com/magazine/archive/2018/12/the-sex
   -recession/573949; "No Sex Please, We're Millennials," *Economist*, May 4, 2019,
   p. 27; and Simon Copland, "The Many Reasons That Young People Are Having
   Less Sex," BBC, May 9, 2017, at http://www.bbc.com/future/story/20170508-the
   -many-reasons-that-people-are-having-less-sex.
4. This desire has also been expanded by women's increased participation in the
   workforce and the advances in reproductive technologies discussed in chapter 4.
5. Cited in Judith Shulevitz, "Why Is Dating in the App Era Such Hard Work?" *Atlantic*, November 2016, pp. 52–54. See also Emily Witt, *Future Sex* (New York:
   Farrar, Straus and Giroux, 2016); and Moira Weigel, *Labor of Love: The Invention of
   Dating* (New York: Farrar, Straus and Giroux, 2016).
6. Abigail Haworth, "Why Have Young People in Japan Stopped Having Sex?" *Guardian*, October 20, 2013, at http://www.theguardian.com/world/2013/oct/20/young
   -people-japan-stopped-having-sex. See also Shiv Malik, "The Dependent Generation: Half Young European Adults Live with their Parents," *Guardian*, March 24,
   2014, at https://www.theguardian.com/society/2014/mar/24/dependent-genera
   tion-half-young-european-adults-live-parents; and Meiji Yasuda Lifestyle Welfare
   Institute, at https://news.mynavi.jp/article/20130401-a041/.
7. One central goal of such matchmaking was to prevent Jewish sons and daughters
   from marrying outside the faith. See Rabbi Maurice Lamm, "Matchmaker,
   Matchmaker, Make Me a Match," Aish.com, January 5, 2002, at http://www.aish
   .com/48960726.html.
8. See Raja Adulrahim, "A Matchmaking Tradition with an Up-to-Date Twist," *Los

Angeles Times, December 26, 2008, at http://articles.latimes.com/2008/dec/26/local/me-biodata26; and Yang Hu, "Marriage of Matching Doors: Marital Sorting on Parental Background in China," *Demographic Journal*, vol. 35, no. 20 (August 31, 2016), pp. 557–80.

9. The true story of one William France, born in 1727. See Geoffrey Castle, "The France Family of Upholsters and Cabinet-Makers," *Furniture History*, vol. 41 (2005), pp. 25–43.

10. The story, of course, is from *Fiddler on the Roof*, which captures this inflection point so poignantly.

11. In Japan, the median age of first marriage in 1950 was 23.6 for women and 26.4 for men. In England, it was respectively 23 and 25. See Fumie Kumagi, "The Life Cycle of the Japanese Family," *Journal of Marriage and Family*, vol. 46, no. 1 (February 1984), pp. 191–204; and "Age and Previous Marital Status at Marriage," Office for National Statistics (2008), at https://www.ons.gov.uk/peoplepopulationandcommunity/birthsdeathsandmarriages/marriagecohabitationandcivilpartnerships/datasets/ageandpreviousmaritalstatusatmarriage.

12. C. Stewart Gillmor, "Stanford, the IBM 650, and the First Trials of Computer Date Matching," *IEEE Annals of the History of Computing*, vol. 29, no. 1 (2007), pp. 74–80.

13. The escorts were not sexual; at this time, women needed male escorts to attend most nighttime social functions. See Marie Hicks, "Computer Love: Replicating Social Order Through Early Computer Dating Systems," *Ada: A Journal of Gender, New Media & Technology*, no. 10 (Fall 2016).

14. See Nick Paumgarten, "Looking for Someone: Sex, Love, and Loneliness on the Internet," *New Yorker*, July 4, 2011, pp. 36–49.

15. See Dan Slater, *Love in the Time of Algorithms* (New York: Current, 2013), pp. 35–36; and Todd Krieger, "Love and Money," *Wired*, September 1, 1995, at https://www.wired.com/1995/09/love-and-money.

16. Slater, *Love in the Time of Algorithms*, pp. 45–46.

17. Kremen sold the company in 1997 for $7 million, reportedly walking away with just a small profit and no girlfriend. See Julia Angwin, "Love's Labor Lost: Online Matchmaker Still Seeks Love, Money," *San Francisco Chronicle*, February 12, 1998, at https://www.sfgate.com/business/article/LOVE-S-LABOR-LOST-Online-matchmaker-still-seeks-3013097.php.

18. "Apple Sells One Million iPhone 3Gs in First Weekend," Apple press release, July 14, 2008, at https://www.apple.com/newsroom/2008/07/14Apple-Sells-One-Million-iPhone-3Gs-in-First-Weekend.

19. "The App Store Turns 10," Apple.com, July 5, 2018, at https://www.apple.com/newsroom/2018/07/app-store-turns-10/.

20. Ibid.

21. Gaby David and Carolina Cambre, "Screened Intimacies: Tinder and the Swipe Logic," *Social Media and Society*, vol. 2, no. 2 (April 2017), p. 3.

22. There is some controversy over Tinder's origins. Rad and Mateen are widely credited as its founders, but others were also involved, including several friends of Rad and Mateen's and an incubator backed by the media conglomerate IAC.

See Shikhar Ghosh et al., "Swipe Right? Dinesh Moorjani and Hatch Labs," Harvard Business School Case N9-818-026; and Steven Bertoni, "Sean Rad Out as Tinder CEO," *Forbes*, November 4, 2014, at https://www.forbes.com/sites/stevenbertoni/2014/11/04/exclusive-sean-rad-out-as-tinder-ceo-inside-the-crazy-saga/#592e708c3ccd.

23. David and Cambre, "Screened Intimacies," p. 3.

24. Issie Lapowsky, "Tinder May Not Be Worth $5B, but It's Way More Valuable Than You Think," *Wired*, April 11, 2014, at https://www.wired.com/2014/04/tinder-valuation.

25. Sindy R. Sumter, Laura Vandenbosch, and Loes Ligtenberg, "Love Me Tinder: Untangling Emerging Adults' Motivations for Using the Dating Application Tinder," *Telematics and Informatics*, vol. 34 (2017), pp. 67–78; and Elisabeth Timmermans and Cédric Courtois, "From Swiping to Casual Sex and/or Committed Relationship: Exploring the Experience of Tinder Users," *The Information Society*, vol. 34, no. 2 (2018), pp. 59–70.

26. Conversation with author, June 4, 2018, New York City.

27. Scientists refer to this as homophily—people's tendency to like those whom they see as similar to themselves. It appears as strong in online attraction as in real life. See A. T. Fiore and J. S. Donath, "Homophily in Online Dating: When Do You Like Someone Like Yourself?" *CHI'05 Extended Abstracts on Human Factors in Computing Systems* (April 2005), pp. 1371–74; and Danaja Maldeniya et al., "The Role of Optimal Distinctiveness and Homophily in Online Dating," *Proceedings of the Eleventh International AAI Conference on Web and Social Media* (May 2017), pp. 616–19.

28. Donna Freitas, an author who surveyed several thousand students about hookup culture, writes, "Hookup culture teaches young people that to become sexually intimate means to become emotionally empty, that in gearing themselves up for sex, they must at the same time drain themselves of feeling." Donna Freitas, *The End of Sex: How Hookup Culture Is Leaving a Generation Unhappy, Sexually Unfulfilled, and Confused About Intimacy* (New York: Basic Books, 2013), p. 11.

29. Here and throughout the personal anecdotes in this chapter, I have changed people's names and minor identifying characteristics.

30. Josué Ortega and Philipp Hergovich, "The Strength of Absent Ties: Social Integration via Online Dating," September 14, 2018, at https://arxiv.org/pdf/1709.10478.pdf; "First Evidence That Online Dating Is Changing the Nature of Society," *MIT Technology Review*, October 10, 2017, at https://www.technologyreview.com/s/609091/first-evidence-that-online-dating-is-changing-the-nature-of-society/; "Interracial Marriage: Who Is Marrying Out?" Pew Research Center, June 12, 2015, at http://www.pewresearch.org/fact-tank/2015/06/12/interracial-marriage-who-is-marrying-out/; and "2011 Census Analysis: What Does the 2011 Census Tell Us About Inter-Ethnic Relationships?" UK Office for National Statistics, July 3, 2014, at https://www.ons.gov.uk/peoplepopulationandcommunity/birthsdeathsandmarriages/marriagecohabitationandcivilpartnerships/articles/whatdoesthe2011censustellusaboutinterethnicrelationships/2014-07-03.

31. Ortega and Hergovich, "The Strength of Absent Ties," p. 8.

32. Cited in "First Evidence That Online Dating Is Changing the Nature of Society."

33. Ortega and Hergovich, "The Strength of Absent Ties."

34. Data are from England and Wales and count both people living with or married to someone from another racial group. See "2011 Census Analysis: What Does the 2011 Census Tell Us About Inter-Ethnic Relationships?"

35. This transition is still underway. Many in India, including those in positions of political power, remain traditionalists when it comes to dating and marriage. See Vidhi Doshi, "Date, Kiss or Marry . . . How Tinder Is Rewriting India's Rules of Engagement," *Guardian*, July 9, 2016, at https://www.theguardian.com/world /2016/jul/09/india-love-revolution-dating-fun-arranged-marriage-apps-tinder; and Nayantara Kilachand, "Mumbai Journal: An Online Dating Revolution, in India," *Wall Street Journal*, June 15, 2012, at https://blogs.wsj.com/indiarealtime /2012/06/15/mumbai-journal-an-online-dating-revolution-in-india/.

36. Quoted in Gardiner Harris, "Websites in India Put a Bit of Choice into Arranged Marriages," *New York Times*, April 24, 2015, at http://www.nytimes.com/2015/04 /26/world/asia/india-arranged-marriages-matrimonial-websites.html.

37. Jialin Li and Anna Lipscomb, "Love on the Cloud: The Rise of Online Dating in China," *US-China Today*, July 17, 2017, at https://china.usc.edu/love-cloud-rise -online-dating-china; and Ben Hubbard, "Young Saudis Find Freedom on Smart-phones," *New York Times*, May 24, 2015, pp. 6, 11.

38. Traditionally, Saudi women are obliged to cover their bodies, and usually their faces, when meeting a potential spouse, and to do so only in the presence of a parent or other chaperone. Reviewing profiles online, therefore, actually gives them the opportunity to learn more about a possible husband than they could during a personal meeting. See Ayman Naji Bajnaid and Tariq Elyas, "Exploring the Phenomena of Online Dating Platforms Versus Saudi Traditional Courtship in the 21st Century," *Digest of Middle East Studies*, vol. 26, no. 1 (2017), pp. 74–96.

39. The classic study here is S. S. Iyengar and M. R. Lepper, "When Choice Is Demotivating: Can One Desire Too Much of a Good Thing?" *Journal of Personality and Social Psychology*, vol. 79, no. 6 (2000), pp. 995–1006. For similar findings in the realm of speed dating, see A. P. Lenton and M. Francesconi, "Too Much of a Good Thing? Variety Is Confusing in Mate Choice," *Biology Letters*, vol. 7 (2011), pp. 528–31.

40. Researchers often refer to this phenomenon as becoming "cognitively over-whelmed." See Eli Finkel et al., "Online Dating: A Critical Analysis from the Perspective of Psychological Science," *Psychological Science in the Public Interest*, vol. 13, no. 1 (2012), esp. p. 33. For a broader analysis, see Barry Schwartz, *The Paradox of Choice: Why More Is Less* (New York: Ecco/HarperCollins, 2004).

41. Private communication with author, June 14, 2018.

42. Davis Webster, "Swiping Right, Staying Put," *New York Times*, May 10, 2015, p. ST6.

43. Quoted in Phoebe Lett, "You Up? College in the Age of Tinder," *New York Times*, February 14, 2018, at https://www.nytimes.com/2018/02/14/opinion/you-up -college-in-the-age-of-tinder.html.

44. See Jean M. Twenge, Ryne A. Sherman, and Brooke E. Wells, "Sexual Inactivity During Young Adulthood Is More Common Among U.S. Millennials and iGen," *Archives of Sexual Behavior*, vol. 46, no. 2 (February 2017), pp. 433–40; Tara Bah-

rampour, "'There Isn't Really Anything Magical About It': Why More Millennials Are Avoiding Sex," *Washington Post*, August 2, 2016, at https://www.washingtonpost .com/local/social-issues/there-isnt-really-anything-magical-about-it-why -more-millennials-are-putting-off-sex/2016/08/02/e7b73d6e-37f4-11e6-8f7c -d4c723a2becb_story.html; and Julian, "Why Are Young People Having So Little Sex?"

45. See Simon Copland, "The Many Reasons That Young People Are Having Less Sex," BBC, May 9, 2017, at http://www.bbc.com/future/story/20170508-the-many -reasons-that-people-are-having-less-sex; and Abigail Haworth, "Why Have Young People in Japan Stopped Having Sex?" *Guardian*, October 20, 2013, at http://www .theguardian.com/world/2013/oct/20/young-people-japan-stopped-having-sex.

46. Quoted in Haworth, "Why Have Young People in Japan Stopped Having Sex?"

47. Ibid.; and Julian, "Why Are Young People Having So Little Sex?"

48. Quoted in Amia Srinivasan, "Does Anyone Have the Right to Sex?" *London Review of Books*, vol. 40, no. 6, March 22, 2018, p. 5.

49. Classic discussions of the asymmetry of sexual power include Catharine MacKinnon, *Women's Lives, Men's Laws* (Cambridge, MA: Harvard University Press, 2005), and Robin Morgan, *The Demon Lover: On the Sexuality of Terrorism* (New York: W. W. Norton, 1989).

50. In one recent study, men matched with only 0.6 percent of the profiles they viewed, compared with 10.5 percent for women. Gareth Tyson et al., "A First Look at User Activity on Tinder," *Proceedings of the 2016 IEEE/ACM International Conference on Advances in Social Networks Analysis and Mining*, 2016 (August 2016), pp. 461–66.

51. Christian Rudder, *Dataclysm* (New York: Broadway Books, 2014), p. 193.

52. Quoted in Mitchell Hobbs, Stephen Owen, and Livia Gerber, "Liquid Love? Dating Apps, Sex, Relationships and the Digital Transformation of Intimacy," *Journal of Sociology*, vol. 53, no. 2 (2017), p. 278.

53. See, for example, Jordan B. Peterson, *12 Rules for Life: An Antidote to Chaos* (Toronto: Random House Canada, 2018).

54. Zygmunt Bauman, *Liquid Love: On the Frailty of Human Bonds* (Cambridge, UK: Polity Press, 2003), p. 65.

55. Quoted in Nancy Jo Sales, "Tinder and the Dawn of the Dating Apocalypse," *Vanity Fair*, September 2015, at http://www.vanityfair.com/culture/2015/08/tinder -hook-up-culture-end-of-dating. His name and some details are disguised in the original.

56. For more on the general phenomenon of Internet addiction, see Liraz Margalit, "Why Are the Candy Crushes of the World Dominating Our Lives?" *Psychology Today*, August 7, 2015, at http://www.psychologytoday.com/us/blog/behind -online-behavior/201508/why-are-the-candy-crushes-the-world-dominating-our -lives.

57. Quoted in Sales, "Tinder and the Dawn of the Dating Apocalypse." His name and some details are disguised in the original.

58. Dopamine also plays a key role in prompting the body to move. People who suffer from Parkinson's disease do not produce sufficient amounts of dopamine. See

O. Hornykiewicz, "The Discovery of Dopamine Deficiency in the Parkinsonian Brain," in Peter Riederer et al., eds., *Parkinson's Disease and Related Disorders* (Vienna: Springer, 2006), pp. 9–15.

59. In men (or at least male prairie voles), a hormone called vasopressin also seems to affect the willingness to bond with a single partner. See Helen E. Fisher, "Brains Do It: Lust, Attraction, and Attachment," *Cerebrum*, January 1, 2000, at http://www.dana.org/Cerebrum/Default.aspx?id=39351; and Krishna G. Seshadri, "The Neuroendocrinology of Love," *Indian Journal of Endocrinology and Metabolism*, vol. 20, no. 4 (July–August 2016), pp. 558–73.

60. The original work in this area was done by B. F. Skinner. See Skinner, *The Behavior of Organisms* (New York: Appleton-Century-Crofts, 1930).

61. Weigel, *Labor of Love*, p. 92.

62. John T. Cacioppo et al., "Marital Satisfaction and Break-Ups Differ Across Online and Off-line Meeting Venues," *Proceedings of the National Academy of Sciences of the United States of America*, vol. 110, no. 25 (June 18, 2013), pp. 10135–40.

63. Quoted in Slater, *Love in the Time of Algorithms*, p. 122. Similar preferences have been demonstrated in the social science laboratory: people who selected a chocolate from an array of six options thought that their choice tasted better than one selected from an array of thirty. See Iyengar and Lepper, "When Choice Is Demotivating"; and Schwartz, *The Paradox of Choice*.

64. Quoted in Sales, "Tinder and the Dawn of the Dating Apocalypse." His name and some details are disguised in the original.

65. See Robert J. Levin, "The Redbook Report on Premarital and Extramarital Sex," *Redbook*, vol. 145, issue 6 (October 1975), pp. 38, 40, 42, 44, 190, 192.

66. For historical data, see National Center for Health Statistics: Vital Statistics of the United States, 7980, vol. 111, Marriage and Divorce. DHHS Publ. No. (PHS) 851103. Public Health Service (Washington, DC: Government Printing Office, 1985).

67. According to one recent study, education-based marriages (in which spouses have similar levels of schooling) are increasing along with inequality. See Robert D. Mare, "Educational Homogeneity in Two Gilded Ages," *Annals of the American Academy of Political and Social Science*, vol. 663, no. 1 (January 2016), pp. 117–39.

68. Olga Abramova et al., "Gender Differences in Online Dating: What Do We Know So Far?" *49th Hawaii International Conference on System Sciences* (January 2016), pp. 3858–67; and Günter J. Hitsch, Ali Hortacsu, and Dan Ariely, "Matching and Sorting in Online Dating," *American Economic Review*, vol. 100, no. 1 (March 2010), pp. 130–63.

69. Some of the most fascinating data have been compiled by Christian Rudder, one of the early partners in OkCupid. In 2009, he began posting a blog called *OkTrends* that drew upon the vast trove of dating data available from the site. The blog posts are available at http://blog.okcupid.com. Other studies include Rudder, *Dataclysm*; Ken-Hou Lin and Jennifer Lundquist, "Mate Selection in Cyberspace: The Intersection of Race, Gender, and Education," *American Journal of Sociology*, vol. 119, no. 1 (July 2013), pp. 183–215; Hitsch, Hortacsu, and Ariely, "Matching and Sorting in Online Dating," pp. 130–63; Günter J. Hitsch, Ali Hortacsu, and Dan

Ariely, "What Makes You Click?—Mate Preferences in Online Dating," *Quantitative Marketing & Economics*, vol. 8 (2010), pp. 393–427; and Jon Birger, *Dateonomics* (New York: Workman Publishing, 2015). For general preferences, see Slater, *Love in the Time of Algorithms*, pp. 81–103.

70. David Wygant, "The Shocking Truth About Tinder Dating!" *Huffington Post*, April 10, 2014, at https://www.huffingtonpost.com/david-wygant/the-shocking -truth-about-_3_b_4967472.html. His anecdotal report is supported by the data: see Abramova et al., "Gender Differences in Online Dating"; and Derek A. Kreager, Shannon E. Cavanagh, John Yen, and Mo Yu, "Where Have All the Good Men Gone? Gendered Interactions in Online Dating," *Journal of Marriage and Family*, vol. 76 (April 2014), pp. 387–410.

71. See Tyson et al., "A First Look at User Activity on Tinder," pp. 461–66; and Abramova et al., "Gender Differences in Online Dating."

72. "No Sex Please, We're Millennials." By comparison, the percentage of women reporting a similar lack of sex has increased by only 8 percent.

73. Pew Research Center, "Millennials in Adulthood: Detached from Institutions, Networked with Friends," March 2014, at https://www.pewsocialtrends.org /2014/03/07/millennials-in-adulthood/; and Steven P. Martin, Nan Marie Astone, and H. Elizabeth Peters, "Fewer Marriages, More Divergence: Marriage Projections for Millennials to Age 40," *Urban Institute*, April 2014, at http://www .urban.org/sites/default/files/publication/22586/413110-Fewer-Marriages-More -Divergence-Marriage-Projections-for-Millennials-to-Age-.PDF.

74. See "Japan Sees Record Population Drop Amid Declining Births," *Nikkei Asian Review*, June 1, 2018, at https://asia.nikkei.com/Economy/Japan-sees-record -population-drop-amid-declining-births.

75. These same redistributive trends are happening more widely as well. See Tyler Cowen, *The Average Is Over* (New York: Dutton, 2013).

76. Herbert Marcuse, *Eros and Civilization: A Philosophical Inquiry into Freud*, 2nd ed. (Boston: Beacon Press, 1966). See also Marcuse, *An Essay on Liberation* (Boston: Beacon Press, 1969).

6. Mad Men

1. See, for example, John Kenneth Galbraith, *The Great Crash: 1929* (Boston: Houghton Mifflin, 1972); Charles P. Kindleberger, *The World in Depression 1929–1939* (Berkeley: University of California Press, 1973); Robert Skidelsky, *Keynes: The Return of the Master* (New York: Public Affairs, 2009); and Luigi Pasinetti and Bertram Schefold, eds., *The Impact of Keynes on Economics in the 20th Century* (London: Edward Elgar, 1999).

2. John Maynard Keynes, "Economic Possibilities for Our Grandchildren" (1930), in *Essays in Persuasion* (New York: Harcourt, Brace, 1932), pp. 358–73.

3. See Rebecca J. Rosen, "Unimate: The Story of George Devol and the First Robotic Arm," *Atlantic*, August 16, 2011, at https://www.theatlantic.com/technology /archive/2011/08/unimate-the-story-of-george-devol-and-the-first-robotic-arm /243716/.

4. Prior to the development of the programmable logic controller (PLC), other, less

fully automated methods of control had been gradually developing. See Mark Walker, Christopher Bissell, and John Monk, "The PLC: A Logical Development," *Measurement and Control*, vol. 43, no. 9 (November 2010), pp. 280–84.

5. There is a considerable literature devoted to tracing this shift and its effects. See, for example, Louis A. Ferman, "The Unmanned Factory and the Community," *Annals of the American Academy of Political and Social Science*, vol. 470 (1983), pp. 136–45; William Serrin, "Impact on Job Market: Divided Predictions," *New York Times*, March 28, 1982, section 2, pp. 39–40; and Martin Ford, *Rise of the Robots: Technology and the Threat of a Jobless Future* (New York: Basic Books, 2015).

6. Jared Yates Sexton, *The Man They Wanted Me to Be: Toxic Masculinity and a Crisis of Our Own Making* (Berkeley: Counterpoint, 2019), p. 23.

7. See Mitra Toossi, "A Century of Change: U.S. Labor Force from 1950 to 2050," *Monthly Labor Review*, vol. 125, no. 5 (May 2002), pp. 15–28. By contrast, in 1950, only one of out of every twenty men of prime working age was unemployed. See Hanna Rosin, *The End of Men: And the Rise of Women* (New York: Riverhead Books, 2012), p. 86. By a different calculation, 31 percent of American men were out of the labor force (either unemployed or not looking for a job) in 2018, up from 14 percent in 1950. See N. Gregory Mankiw, "Why Aren't More Men Working?" *New York Times*, June 15, 2018, at https://www.nytimes.com/2018/06/15/business/men-unemployment-jobs.html?smid=nytcore-ios-share.

8. "Men Adrift," *Economist*, May 30, 2015, p. 23.

9. Data from the World Bank, at https://data.worldbank.org/indicator/SL.UEM.TOTL.MA.ZS?locations=US-IT.

10. Steven Greenhouse, "Is Trump Really Pro-Worker?" *New York Times*, Sunday Review, September 3, 2017, p. 2; Stanley Lebergott, "Labor Force Participation and Employments, 1800–1960," in *Output, Employment, and Production in the United States after 1800*, ed. Dorothy Brady (Washington, DC: National Bureau of Economic Research, 1966), pp. 117–204, at http://ww.nber.org/chapters/c1567.pdf; and data from the U.S. Bureau of Labor Statistics, at https://www.bls.gov/oes/current/oes472221.htm.

11. Brian Wang, "Foxconn Reaches 40,000 Robots of Original 1 Million Robot Automation Goal," Next Big Future, October 13, 2016, at https://www.nextbigfuture.com/2016/10/foxconn-reaches-40000-robots-of.html.

12. See Claudia Goldin, *Understanding the Gender Gap: An Economic History of American Women* (New York: Oxford University Press, 1990).

13. Catherine Rampell, "As Layoffs Surge, Women May Pass Men in Job Force," *New York Times*, February 5, 2009, p. A1.

14. Kirkpatrick Sale, *Rebels Against the Future* (London: Quartet Books, 1995).

15. Cited in Erik Brynjolfsson and Andrew McAfee, *The Second Machine Age* (New York: W. W. Norton, 2014), p. 143.

16. Ibid., p. 130; Olivia LaVecchia and Stacy Mitchell, "Amazon's Stranglehold: How the Company's Tightening Grip Is Stifling Competition, Eroding Jobs, and Threatening Communities," Institute for Local Self-Reliance, November 2016, at https://ilsr.org/wp-content/uploads/2016/11/ILSR_AmazonReport_final.pdf; and David Rotman, "How Technology Is Destroying Jobs," *MIT Technology Review*, June 12, 2013,

at https://www.technologyreview.com/s/515926/how-technology-is-destroying
-jobs/.

17. Lawrence H. Summers, "Economic Possibilities for Our Children," *NBER Reporter*, no. 4 (2013), at http://www.nber.org/reporter/2013number4/2013no4.pdf. Data give further evidence of this point, showing that the global labor share has significantly declined since the 1980s, as a result of information-based technologies. See Loukas Karabarbounis and Brent Neiman, "The Global Decline of the Labor Share," National Bureau of Economic Research, working paper no. 19136, 2013. For an argument that humans will still manage to find new sources of employment, see Erik Brynjolfsson and Andrew McAfee, "Will Humans Go the Way of Horses?" *Foreign Affairs*, vol. 94, no. 4 (July–August 2015), pp. 8–14.

18. Economists generally refer to this trend as "skill-biased technical change." For evidence of its existence and power, see David H. Autor, Lawrence F. Katz, and Alan B. Krueger, "Computing Inequality: Have Computers Changed the Labor Market?" National Bureau of Economic Research, working paper no. 5956, March 1997, at http://www.nber.org/papers/w5956; F. Levy and R. J. Murnane, *The New Division of Labor: How Computers Are Creating the Next Job Market* (Princeton: Princeton University Press, 2012); and Daron Acemoglu and David Autor, "Skills, Tasks and Technologies: Implications for Employment and Earnings," National Bureau of Economic Research, working paper no. 16082, June 2010, at http://www.nber.org/papers/w16082.

19. Their employment trends have also been much stronger since the end of the 2008 recession. See, for instance, Michelle Jamrisko, "Women Charge Past Men in US Job Market as Economy Lumbers On," *Bloomberg*, August 16, 2017, at https://www.bloomberg.com/news/articles/2017-08-16/women-charge-past-men-in-u-s-job-market-as-economy-lumbers-on.

20. On men's persistent reluctance to enter nursing, one of the economy's most stable and well-paid professions, see Karin Fischer, "The Stubborn Stigma of the Male Nurse," *Chronicle of Higher Education*, September 1, 2017, pp. A14–18.

21. As of 2010, the majority of pharmacists in the United States were female. They enjoyed a considerably smaller wage gap vis-à-vis men than was the case in most other industries and a reduced penalty for part-time work. See Claudia Goldin and Lawrence F. Katz, "A Most Egalitarian Profession: Pharmacy and the Evolution of a Family-Friendly Occupation," *Journal of Labor Economics*, vol. 34, no. 3 (2016), pp. 705–46.

22. See Derek Thompson, "It's Not Just a Recession. It's a Mancession!" *Atlantic*, July 9, 2009, at https://www.theatlantice.com/business/archive/2009/07/its-not-just-a-recession-its-a-mancesdsion/20991/.

23. There is considerable, and worrisome, evidence that the gap between the highest-paid and lowest-paid workers is continuing to widen, leading to smaller and smaller numbers of people controlling ever-larger fractions of the world's overall wealth. See, for instance, Brynjolfsson and McAfee, *The Second Machine Age*, pp. 148–85; and Robert Frank and Philip Cook, *The Winner-Take-All Society* (New York: Penguin Books, 1996).

24. Fischer, "The Stubborn Stigma of the Male Nurse." See also "The Weaker Sex," *Economist*, May 30, 2015, p. 11; and Jeff Guo, "The Serious Reasons Boys Do Worse Than Girls," *Washington Post*, January 28, 2016, at https://www.washingtonpost .com/news/wonk/wp/2016/01/28/the-serious-reason-boys-do-worse-than-girls/ ?noredirect=on&utm_term=.be67892a951b.

25. Arlie Russell Hochschild makes this point poignantly in the afterword to her in-depth analysis of a working-class community in Louisiana. See Arlie Russell Hochschild, *Strangers in Their Own Land: Anger and Mourning on the American Right* (New York: The New Press, 2018), pp. 260–66.

26. CBS News, "The Town Maytag Left Behind," October 18, 2007, at https://www .cbsnews.com/news/the-town-maytag-left-behind/.

27. Louis Uchitelle, "Is There (Middle Class) Life After Maytag?" *New York Times*, August 26, 2007, at https://www.nytimes.com/2007/08/26/business/yourmoney /26maytag.html?mcubz=1.

28. CBS News, "The Town Maytag Left Behind."

29. Ibid.

30. Hanna Rosin, "Who Wears the Pants in This Economy?" *New York Times Magazine*, September 2, 2012, p. MM22.

31. Ibid.; and Hanna Rosin, *The End of Men: And the Rise of Women* (New York: Riverhead Books, 2012).

32. "Impact of the Economic Crisis Explained," Eurostat: Statistics Explained, Archive, July 2009, at http://ec.europa.eu/eurostat/statistics-explained/index.php/Archive: Impact_of_the_economic_crisis_on_unemployment.

33. "International Comparisons of Annual Labor Force Statistics, 1970–2012," Bureau of Labor Statistics, June 7, 2013, at https://www.bls.gov/fls/flscomparelf /lfcompendium.pdf.

34. Bureau of Labor Statistics, reported in Claire Cain Miller, "Why Men Don't Want Jobs Mostly Done by Women," *New York Times*, January 5, 2017, p. A3.

35. For a history of the development of gender studies, see Cordelia Fine, *Delusions of Gender: How Our Minds, Society, and Neurosexism Create Difference* (New York: W. W. Norton, 2010); and Anne Fausto-Sterling, *Sexing the Body: Gender, Politics, and the Construction of Sexuality* (New York: Basic Books, 2000).

36. While most public universities in the United States had been open to women since their inception, many of the most elite private universities—including Harvard, Yale, and Princeton—started admitting women to their undergraduate colleges only after 1969. Prior to that point, women who sought to be educated at these schools were directed instead to their "sisters," all-women's colleges like Radcliffe and Barnard. Similarly, Oxford and Cambridge had begun accepting women as early as 1920, but only to a handful of female colleges.

37. See, for instance, Sheryl Sandberg's *Lean In* (New York: Knopf, 2013), currently the most well-known (and controversial) voice for corporate feminism. Anne-Marie Slaughter's *Unfinished Business* (New York: Random House, 2016) focuses much more on issues of pay equity and the burden of what she calls care work—although Slaughter is generally also referred to by critics as a corporate feminist. For more on the issues of pay and gender equity, see Heather Boushey, *Finding*

Time: The Economics of Work-Life Conflict (Cambridge, MA: Harvard University Press, 2016).

38. Debora L. Spar, "Why the Woman Who 'Has It All' Doesn't Really Exist," *Glamour*, August 14, 2013, at https://www.glamour.com/story/why-women-cant-have -it-all-according-to-barnard-college-president-debora-l-spar.

39. In *Stiffed*, a relatively early account of men's shifting roles, Susan Faludi argued poignantly that men at the end of the twentieth century were falling into a status "oddly similar to that of women at mid-century." See Susan Faludi, *Stiffed: The Betrayal of the American Man* (New York: William Morrow, 1999), p. 40.

40. See Victor J. Seidler, *Rediscovering Masculinity: Reason, Language and Sexuality* (London: Routledge, 1989).

41. Or, as he writes, "We can re-arrange difference only if we contest dominance." See R. W. Connell, "Gender Politics for Men," *International Journal of Sociology and Social Policy*, vol. 17, no. 1/2 (1997), pp. 62–77; quote appears on page 67. Connell is a trans woman, who completed her transition in 2012. This article was written when she still presented as male.

42. For a devastating critique of this form of masculinity and its effects, see Sexton, *The Man They Wanted Me to Be*. For other academic accounts, see R. W. Connell, *Masculinities* (Berkeley: University of California Press, 1995); C. J. Pascoe and Tristan Bridges, *Exploring Masculinities: Identity, Inequality, Continuity and Change* (Oxford: Oxford University Press, 2016); Richard Howson, *Challenging Hegemonic Masculinity* (New York: Routledge, 2006); and James W. Messerschmidt, *Hegemonic Masculinity: Formulation, Reformulation, and Amplification* (Lanham, MD: Rowman & Littlefield, 2018). For a celebratory account, see Harvey Mansfield, *Manliness* (New Haven: Yale University Press, 2006).

43. Sociologist Michael Kimmel refers to this as a "masculinity bonus." See Michael Kimmel, *Angry White Men: American Masculinity at the End of an Era* (New York: Bold Type Books, 2019), p. 8.

44. Victor Seidler, one of the few sociologists who has been examining masculinity for some time, argues that the very pervasiveness of male power has made it more difficult for men to examine different male experiences and masculinities. See Seidler, *Rediscovering Masculinity*.

45. See, for example, Stephanie Russell-Kraft, "The Rise of the Male Supremacist," *New Republic*, April 4, 2018, at https://newrepublic.com/article/147744/rise-male -supremacist-groups; 2018 Anti-Defamation League Report, *When Women Are the Enemy: The Intersection of Misogyny and White Supremacy*, at https://www.adl.org /resources/reports/when-women-are-the-enemy-the-intersection-of-misogyny -and-white-supremacy#the-alt-right-has-a-woman-problem; and Sexton, *The Man They Wanted Me to Be*, pp. 120–27. The Southern Poverty Law Center, which monitors all sorts of hate groups, made this link in 2012, adding men's rights groups to its annual survey. See Southern Poverty Law Center, "Misogyny: The Sites," *Intelligence Report*, no. 145 (Spring 2012), at https://www.splcenter.org /fighting-hate/intelligence-report/2012/misogyny-sites; and, more generally, Michael Kimmel, *Angry White Men*, pp. 99–134.

46. Sexton, *The Man They Wanted Me to Be*, p. 5.

47. Betty Friedan, *The Feminine Mystique* (New York: Penguin Books, 2010), p. 202.
48. Rosin, *The End of Men*, pp. 57–58.
49. For a review of the academic literature on what has been called "hegemonic masculinity," see R. W. Connell and James W. Messerschmidt, "Hegemonic Masculinity: Rethinking the Concept," *Gender and Society*, vol. 19, no. 6 (December 2005), pp. 829–59.
50. Marc Tracy, "Here Come the Daddy Wars," *New Republic*, June 14, 2013, at https://newrepublic.com/article/113490/daddy-wars-will-be-mommy-wars-men.
51. James Houghton, "Foreword," in James Houghton, Larry Bean, and Tom Matlack, eds., *The Good Men Project* (Boston: The Good Men Foundation, 2009), p. 2.
52. See Olga Khazan, "Middle-Aged White Americans Are Dying of Despair," *Atlantic*, November 2015, at https://www.theatlantic.com/health/archive/2015/11/boomers-deaths-pnas/413971/. For the original research, see Anne Case and Angus Deaton, "Rising Morbidity and Mortality in Midlife Among White Non-Hispanic Americas in the 21st Century," *Proceedings of the National Academy of Sciences*, vol. 112, no. 49 (December 8, 2015), pp. 15078–83. In addition to the overall trends, which have affected both men and women, suicide rates among middle-aged men in the United States have risen significantly since 2000, with particular increases among unmarried men and those without a college education. See Julie Phillips et al., "Understanding Recent Changes in Suicide Rates Among the Middle-Aged: Period or Cohort Effects," *Public Health Reports* 125 (September 2010), pp. 680–88; Stephen Rodrik, "All-American Despair," *Rolling Stone*, May 30, 2019, at https://www.rollingstone.com/culture/culture-features/suicide-rate-america-white-men-841576/; and Sumathi Reddy, "The Mystery Around Middle-Age Suicides," *Wall Street Journal*, June 14, 2018, at https://www.wsj.com/articles/the-mystery-around-middle-age-suicides-1528948801, accessed November 2019. Finally, while rates of death from drug overdoses are rising more quickly for women than for men, men still die from overdoses at far higher rates than women. See Laura Joszt, "CDC Data: Life Expectancy Decreases as Deaths from Suicide, Drug Overdose Increase," *In Focus Blog*, November 30, 2018, at https://www.ajmc.com/focus-of-the-week/cdc-data-life-expectancy-decreases-as-deaths-from-suicide-drug-overdose-increase.
53. The phrase "nasty woman" rose to the fore in a 2016 debate between candidates Donald Trump and Hillary Clinton, when Trump, in the middle of a comment by Secretary Clinton, took to the microphone to declare, "Such a nasty woman." In the wake of the debate, "nasty women" became a rallying cry for many of Trump's opponents.
54. For discussions of gay masculinities, see R. W. Connell, "A Very Straight Gay: Masculinity, Homosexual Experience, and the Dynamics of Gender," *American Sociological Review*, vol. 57, no. 6 (December 1993), pp. 735–51. For more general reviews of masculinity and its variants, see Connell, *Masculinities*; M. C. Gutmann, *The Meanings of Macho: Being a Man in Mexico City* (Berkeley: University of California Press, 1996); and the essays in M. S. Kimmel, ed., *Changing Men: New Directions in Research on Men and Masculinities* (Newbury Park, CA: Sage, 1987).
55. Sadly, he and his brethren have also frequently been accused of massively sexist

behavior—one of the explanations given for the fact that their industries remain so decidedly male-dominated. See Liza Mundy, "Why Is Silicon Valley So Awful to Women?" *Atlantic*, April 2017, at https://www.theatlantic.com/magazine /archive/2017/04/why-is-silicon-valley-so-awful-to-women/517788; and Olga Khazan, "The Sexism of Startup Land," *Atlantic*, March 12, 2015, at https://www.the atlantic.com/business/archive/2015/03/the-sexism-of-startup-land/387184.

56. See Dan Lyons, "In Silicon Valley, Working 9 to 5 Is for Losers," *New York Times*, Sunday Review, September 3, 2017, p. SR2; and Claire Cain Miller and Nick Bilton, "A Foot in the Door in Silicon Valley," *New York Times*, February 5, 2012, p. ST1.

57. See Lyons, "In Silicon Valley, Working 9 to 5 Is for Losers."

58. Matt Oscamou, quoted in "More Than 40 Work-Life Balance Tips from Dad Entrepreneurs," Launch Grow Joy, at https://www.launchgrowjoy.com/more-than -40-work-life-balance-tips-from-dad-entrepreneurs/.

59. In a 2011 study, 60 percent of fathers in dual-earner households reported experiencing work-life conflict, compared with just 35 percent in 1977. See Joan C. Williams, "The Daddy Dilemma: Why Men Face a 'Flexibility Stigma' at Work," *Washington Post*, February 11, 2013, at https://www.washingtonpost.com/national/on -leadership/the-daddy-dilemma-why-men-face-a-flexibility-stigma-at-work/2013/02 /11/58350f43-7462-11e2-aa12-e6cf1d31106b_story.html?utm_term=.7dde32c6b17c.

60. Patrick Strickland, "Alt-Right Rally: Charlottesville Braces for Violence," Al Jazeera, August 11, 2017, at http://www.aljazeera.com/indepth/features/2017/08/alt -rally-charlottesville-braces-violence-170810073156023.html.

61. See May Bulman, "Brexit: People Voted to Leave EU Because They Feared Immigration, Major Survey Finds," *Independent*, June 28, 2017, at http://www.independent .co.uk/news/uk/home-news/brexit-latest-news-leave-eu-immigration-main -reason-european-union-survey-a7811651.html.

7. Trans*itions

1. For statistics, see S. E. James, J. L. Herman, S. Rankin, M. Keisling, L. Mottet, and M. Anafi, *The Report of the 2015 U.S. Transgender Survey* (Washington, DC: National Center for Transgender Equality, 2016); and Laura Edwards-Leeper and Norman P. Spack, "Psychological Evaluation and Medical Treatment of Transgender Youth in an Interdisciplinary 'Gender Managements Service' (GeMS) in a Major Pediatric Center," *Journal of Homosexuality*, vol. 59, no. 3 (2012), pp. 321–36.

2. Dhananjay Mahapatra, "Supreme Court Recognizes Transgenders as 'Third Gender,'" *The Times of India*, April 15, 2014, at https://timesofindia.indiatimes.com /india/Supreme-Court-recognizes-transgenders-as-third-gender/articleshow /33767900.cms; and Will Oremus, "Here Are All the Different Genders You Can Be on Facebook," *Slate*, February 13, 2014, at https://slate.com/technology/2014/02 /facebook-custom-gender-options-here-are-all-56-custom-options.html.

3. Plato, *Symposium*, *Plato in Twelve Volumes: Lysis, Symposium, Gorgias*, trans. W. R. M. Lamb (Cambridge, MA: Harvard University Press, 1925), 189c–193, 189e–193b. See also Michael Groneberg, "Myth and Science Around Gender and Sexuality: Eros and the Three Sexes in Plato's *Symposium*," *Diogenes*, vol. 52, no. 4 (November 2005), pp. 39–49.

4. See Gerhard J. Newerla, "The History of the Discovery and Isolation of the Female Sex Hormones," *New England Journal of Medicine*, vol. 230, no. 20 (May 18, 1944), pp. 595–604; and Ellen Goldberg, *The Lord Who Is Half Woman: Ardhanaris-vara in Indian and Feminist Perspective* (Albany: State University of New York Press, 2002).

5. M. Michelraj, "Historical Evolution of Transgender Community in India," *Asian Review of Social Sciences*, vol. 4, no. 1 (2015), pp. 17–19; and Jessie Guy-Ryan, "In Indonesia, Non-Binary Gender Is a Centuries-Old Idea," *Atlas Obscura*, June 18, 2016, at https://www.atlasobscura.com/articles/in-indonesia-nonbinary-gender-is-a-centuriesold-idea.

6. See, for example, Walter Stevenson, "The Rise of Eunuchs in Greco-Roman Antiquity," *Journal of the History of Sexuality*, vol. 5, no. 4 (April 1995), pp. 495–511; and Walter L. Williams, "The 'Two-Spirit' People of Indigenous North Americans," *Guardian*, October 11, 2010, at https://www.theguardian.com/music/2010/oct/11/two-spirit-people-north-america.

7. See "15 Notable Ambiguous Genitalia Statistics," Health Research Funding, at https://healthresearchfunding.org/15-notable-ambiguous-genitalia-statistics.

8. Deborah Rudacille, *The Riddle of Gender: Science, Activism, and Transgender Rights* (New York: Pantheon Books, 2005), p. 3. See also Vern L. Bullough and Bonnie Bullough, *Cross Dressing, Sex, and Gender* (Philadelphia: University of Pennsylvania Press, 1993), Part 1, "Cultural and Historical Background: Cross Dressing in Perspective," p. 5.

9. See Genny Beemyn, "Transgender History in the United States," in *Trans Bodies, Trans Selves*, ed. Laura Erickson-Schroth (Oxford: Oxford University Press, 2014), pp. 501–32; and I. Bennett Capers, "Cross Dressing and the Criminal," *Yale Journal of Law & the Humanities*, vol. 20, issue 1 (Winter 2008), pp. 1–30.

10. Quoted in Magnus Hirschfeld, *Transvestites*, trans. Michael A. Lombardi-Nash (New York: Prometheus Books, 1991), pp. 28–29.

11. See, for instance, Frederic G. Worden and James T. Marsh, "Psychological Factors in Men Seeking Sex Transformation," *Journal of the American Medical Association*, vol. 157, no. 15 (April 9, 1955), pp. 1292–98.

12. For an excellent overview of how medicine came to influence and regulate sexuality in general and transgender individuals in particular, see Susan Stryker, *Transgender History*, 2nd ed. (New York: Seal Press, 2017), pp. 45–77.

13. Ibid., pp. 52–53.

14. Ibid.

15. See David C. Lindberg, *The Beginnings of Western Science*, 2nd ed. (Chicago: University of Chicago Press, 2007), p. 336.

16. For a review of this evolution, see ibid.; and Albert Q. Maisel, *The Hormone Quest* (New York: Random House, 1965).

17. Quoted in Jamshed R. Tata, "One Hundred Years of Hormones," *EMBO Reports*, vol. 6, no. 2 (2005), p. 490.

18. Ernest Starling, "The Croonian Lectures. I. On the Chemical Correlation of the Functions of the Body," *The Lancet*, vol. 166, no. 4275 (August 5, 1905), pp. 339–41.

19. Whatever that might mean in this context. Quoted in John Money, "The Genea-

logical Descent of Sexual Psychoneuroendocrinology from Sex and Health Theory: The Eighteenth to the Twentieth Centuries," *Psychoneuroendocrinology*, vol. 8, no. 4 (1983), p. 391.

20. This researcher, Arnold A. Berthold, also transplanted a testis into a formerly neutered bird (a capon) and concluded that the transplant restored the capon to its precastration state. See Money, "The Genealogical Descent of Sexual Psychoneuroendocrinology from Sex and Health Theory," p. 392.

21. History is unclear as to whether the extract came from dogs or guinea pigs. See Money, "The Genealogical Descent of Sexual Psychoneuroendocrinology from Sex and Health Theory," p. 393.

22. His name was G. R. Murray, cited in ibid., p. 392.

23. Throughout the developmental process, the brain and central nervous systems also play critical roles, interacting with and responding to the relevant hormones. For a more complete description, see Cordelia Fine, *Delusions of Gender: The Real Science Behind Sex Differences* (New York: W. W. Norton, 2010); and Stuart Tobet et al., "Brain Sex Differences and Hormone Influences," *Journal of Neuroendocrinology*, vol. 21, no. 4 (April 2009), pp. 387–92. Much of the scientific debate in Steinach's time revolved around the relative roles of hormones and the central nervous system. See Per Södersten et al., "Eugen Steinach: The First Neuroendocrinologist," *Endocrinology*, vol. 155, no. 3 (March 2014), pp. 690–91; C. Sengoopta, *The Most Secret Quintessence of Life: Sex, Glands, and Hormones, 1850–1950* (Chicago: University of Chicago Press, 2006); and Frank A. Beach, "Historical Origins of Modern Research on Hormones and Behavior," *Hormones and Behavior*, vol. 15, no. 4 (December 1981), pp. 325–76.

24. Quoted in Joanne Meyerowitz, *How Sex Changed: A History of Transsexuality in the United States* (Cambridge, MA: Harvard University Press, 2004), p. 17. For Steinach's work, see Eugen Steinach, "Willkürliche Umwandlung von Säugetier-Männchen in Tiere mit ausgeprägt weiblichen Geschlechtscharakteren und weiblicher Psych. [Arbitrary Transformation of Mammalian Males in Animals with Pronounced Female Sexual Characteristics and Female Psyche]," *Pflügers Archiv European Journal of Physiology* 144 (1912), pp. 71–108; and Steinach, "Feminierung von Männchen und Maskulierung von Weibchen [Feminization of Males and Masculinization of Females], *Zentralblatt für Physiologie*, vol. 27 (1913), pp. 717–23. For a review of his work, see Chandak Sengoopta, "Glandular Politics: Experimental Biology, Clinical Medicine, and Homosexual Emancipation in Fin-de-Siecle Central Europe," *Isis*, vol. 89, no. 3 (September 1998), pp. 445–73. For subsequent, and supporting, work along these lines, see Charles H. Phoenix et al., "Organizing Action of Prenatally Administered Testosterone Propionate on the Tissues Mediating Mating Behavior in the Female Guinea Pig," *Endocrinology*, vol. 65, no. 3 (September 1, 1959), pp. 369–82.

25. It is crucial to note that Steinach did this work before the science of genetics had even really begun.

26. Eugen Steinach, *Sex and Life* (London: Faber and Faber, 1940), pp. 7. Quoted in Meyerowitz, *How Sex Changed: A History of Transsexuality in the United States*, p. 27.

27. Meyerowitz, *How Sex Changed*, p. 27.

28. For a full recounting of the role of synthetic hormones in the pill's creation, see Bernard Asbell, *The Pill: A Biography of the Drug That Changed the World* (New York: Random House, 1995).

29. Södersten et al., "Eugen Steinach: The First Neuroendocrinologist," p. 689.

30. His one partial foray into work on humans did not end well. Along with others of his generation, Steinach came to believe that aging men could be rejuvenated by vasectomy, and he performed such operations a number of times. The public ridicule that eventually surrounded the "Steinach operation" may have contributed to his lack of a Nobel Prize. See, for instance, Sengoopta, "Glandular Politics," pp. 457–58; and Arnold Kahn, "Regaining Lost Youth: The Controversial and Colorful Beginnings of Hormone Replacement Therapy in Aging," *Journal of Gerontology*, series A, vol. 60, no. 2 (February 2005), pp. 142–47.

31. For more on Hirschfeld's fascinating life and work, see Rudacille, *The Riddle of Gender*, pp. 30–61; Charlotte Wolff, *Magnus Hirschfeld: A Portrait of a Pioneer in Sexology* (New York: Quartet Books, 1986); and Sengoopta, "Glandular Politics," pp. 445–73.

32. The operations were reported in Felix Abraham, "Genitalumwandlung an Zwei Männlichen Transvestiten [Genital Reassignment on Two Male Transvestites]," *Zeitschrift für Sexualwissenschaft und Sexualpolitik*, vol. 18 (September 10, 1931), pp. 223–26, reprinted in *International Journal of Transgenderism*, vol. 2, no. 1 (January–March 1998), at https://editions-ismael.com/fr/1931-felix-abraham -genital-reassignment-on-two-male-transvestites/.

33. Ludwig Lenz, *Memoirs of a Sexologist*, quoted in Meyerowitz, *How Sex Changed*, p. 20.

34. Lili Elbe, quoted in Niels Hoyer, ed., *Man into Woman: An Authentic Record of Change* (New York: E. P. Dutton, 1933), p. 250.

35. Ibid., p. 275.

36. Hirschfeld would die in exile in France, two years later.

37. The central figure in the United States was Dr. Henry Benjamin. For descriptions of him and his work, see Harry Benjamin, *The Transsexual Phenomenon* (New York: Julian Press, 1966); and Richard Ekins, "Science, Politics and Clinical Intervention: Harry Benjamin, Transsexualism and the Problem of Heteronormativity," *Sexualities*, vol. 8, no. 3 (July 2005), pp. 306–28.

38. In the United States, the most famous was Christine Jorgensen, a "blond bombshell" whose sex-change surgeries captivated the U.S. media in 1952. See Christine Jorgensen, *Christine Jorgensen: A Personal Autobiography* (San Francisco: Cleis Press, 2000); Meyerowitz, *How Sex Changed*, pp. 51–97; and Rudacille, *The Riddle of Gender*, pp. 62–101. Other highly publicized cases included the tennis player Renée Richards (née Richard Raskind), the activist Virginia Prince (née Arnold Lowman), and the writer Jan Morris (née James Morris).

39. For more on this possibility, see Dick F. Swaab and Alicia Garcia-Falgueras, "Sexual Differentiation of the Human Brain in Relation to Gender Identity and Sexual Orientation," *Functional Neurology*, vol. 24, no. 1 (2009), pp 17–28. For a more general description of Money's work and theories, see John Money and

Clay Primrose, "Sexual Dimorphism and Dissociation in the Psychology of Male Transsexuals," *Journal of Nervous and Mental Disease*, vol. 147, no. 5 (1968), p. 481.

40. J. Money, J. G. Hampson, and J. L. Hampson, "Imprinting and the Establishment of Gender Role," *Archives of Neurology and Psychiatry*, vol. 77 (1957), pp. 333–36. See also Richard Green and John Money, "Incongruous Gender Role: Nongenital Manifestations in Prepubertal Boys," *Journal of Nervous and Mental Disease*, vol. 131, no. 2 (August 1960), pp. 160–68.

41. The quote is from David Reimer, who was treated by Money after a badly failed circumcision and subsequently raised as a girl. As an adult, David eventually resumed his male identity and underwent medical procedures to reverse Money's original work. He committed suicide at the age of thirty-eight. See Elaine Woo, "David Reimer, 38: After Botched Surgery, He Was Raised as a Girl in Gender Experiment," *Los Angeles Times*, May 13, 2004, at http://articles.latimes.com/2004/may/13/local/me-reimer13/2. Also see John Colapinto, *As Nature Made Him: The Boy Who Was Raised as a Girl* (New York: Harper Perennial, 2006).

42. Jane E. Brody, "Benefits of Transsexual Surgery Disputed as Leading Hospital Halts the Procedure," *New York Times*, October 2, 1979, pp. C1, C3; Vernon A. Rosario, "Studs, Stems, and Fishy Boys: Adolescent Gender Variance and the Slippery Diagnosis of Transsexuality," in Chantal Zabus and David Coad, eds., *Transgender Experience: Place, Ethnicity, and Visibility* (New York: Routledge, 2014), p. 54; and L. M. Lothstein, "Sex Reassignment Surgery: Historical, Bioethical, and Theoretical Issues," *American Journal of Psychiatry*, vol. 139, no. 4 (April 1982), pp. 417–26.

43. For more on this work, see P. T. Cohen-Kettenis and S. H. M. van Goozen, "Pubertal Delay as an Aid in Diagnosis and Treatment of a Transsexual Adolescent," *European Child & Adolescent Psychiatry*, vol. 7 (1998), pp. 246–48; L. J. G. Gooren and H. Delemarre-van de Wall, "The Feasibility of Endocrine Interventions in Juvenile Transsexuals," *Journal of Psychology and Human Sexuality*, vol. 8 (1996), pp. 69–74; Annelou L. C. DeVries and Peggy T. Cohen-Kettenis, "Clinical Management of Gender Dysphoria in Children and Adolescents," *Journal of Homosexuality*, vol. 59, no. 3 (2012), pp. 301–20; and Daniel E. Shumer, Natalie J. Nokoff, and Norman P. Spack, "Advances in the Care of Transgender Children and Adolescents," *Advances in Pediatrics*, vol. 63, no. 1 (August 2016), pp. 79–102.

44. Sabrina Rubin, "About a Girl: Coy Mathis' Fight to Change Gender," *Rolling Stone*, October 28, 2013, at https://www.rollingstone.com/culture/culture-news/about-a-girl-coy-mathis-fight-to-change-gender-64264/.

45. Before the case could be heard, the newly elected Trump administration rescinded a Department of Education policy that undergirded the case. See https://www.aclu.org/cases/gg-v-gloucester-county-school-board.

46. Not all liberal voices have historically sided with the transgender cause. In fact, there was for many years an intense fight within the feminist community, with some factions arguing passionately that trans women (MTFs) were neither feminists nor women, and that indeed they were men who "rape women's bodies by reducing the real female form to an artifact." See Janice Raymond, *The Transsexual Empire: The Making of the She-Male* (Boston: Beacon, 1979), p. 104.

47. "Patrick Signals His Support for a Statewide Bathroom Bill," Dan Patrick: Texas Lieutenant Governor, April 26, 2006, at https://www.danpatrick.org/patrick-signals-his-support-for-a-statewide-bathroom-bill/.

48. Tal Kopan and Eugene Scott, "North Carolina Governor Signs Controversial Transgender Bill," CNN, March 24, 2016, at https://www.cnn.com/2016/03/23/politics/north-carolina-gender-bathrooms-bill/.

49. Data from S. E. James et al., "The Report of the 2015 U.S. Transgender Survey" (Washington, DC: National Center for Transgender Equality, 2016), at https://transequality.org/sites/default/files/docs/usts/USTS-Full-Report-Dec17.pdf.

50. See Jennifer Finney Boylan, *She's Not There: A Life in Two Genders* (New York: Broadway Books, 2013).

51. Jennifer Finney Boylan, "My Gay Agenda," *New York Times*, July 24, 2017, p. A19.

52. Boylan, *She's Not There*, p. 21.

53. Ceylan Yeginsu, "Drug Regimen Is Said to Enable Transgender Breast-Feeding," *New York Times*, February 16, 2018, p. A9.

54. "Global Attitudes Towards Transgender People," Ipsos Public Affairs, January 2018, at https://www.ipsos.com/sites/default/files/ct/news/documents/2018-01/ipsos_report-transgender_global_data-2018.pdf.

55. "Accelerating Acceptance," GLAAD, 2017, at https://www.glaad.org/files/aa/2017_GLAAD_Accelerating_Acceptance.pdf, cited in Jared Yates Sexton, *The Man They Wanted Me to Be: Toxic Masculinity and a Crisis of Our Own Making* (Berkeley: Counterpoint, 2019), p. 239; see also Phillip L. Hammack, "The Future Is Non-Binary, and Teens Are Leading the Way," *Pacific Standard*, April 8, 2019, at https://psmag.com/ideas/gen-z-the-future-is-non-binary. For the complete report, see Kimberly Lawsom, "27 Percent of California Teens Are Gender Nonconforming," Williams Institute, UCLA School of Law, December 14, 2017, at https://williamsinstitute.law.ucla.edu/press/in-the-news/27-percent-california-teens-gender-nonconforming/.

56. The leading theorist in this area is Judith Butler, who argues that gender itself is constructed and defined by the ways in which individuals perform. "Gender," she argues, "is manufactured through a sustained set of acts, posited through the gendered stylization of the body." Judith Butler, *Gender Trouble: Feminism and the Subversion of Identity* (New York: Routledge, 1999), p. xv. See also Sandy Stone, "The 'Empire' Strikes Back: A Posttransexual Manifesto," *Camera Obscura*, vol. 10, no. 2 (May 1992), pp. 150–76; and Riki Anne Wilchins, *Read My Lips: Sexual Subversion and the End of Gender* (Ithaca: Firebrand, 1997).

57. Exceptions include Bernice L. Hausman, *Changing Sex* (Durham, NC: Duke University Press, 1995); and Eve Shapiro, *Gender Circuits* (New York: Routledge, 2015).

58. For an argument along these lines, see Sengoopta, "Glandular Politics," p. 470.

8. Cuddling with Robots

1. This quote appeared as the epigraph for Mary Shelley's first edition of *Frankenstein*. It is from John Milton's *Paradise Lost*, book X, lines 743–45. See Mary Shelley, *Frankenstein, or The Modern Prometheus*, vol. 1 (London: Lackington, Hughes, Harding, Mayor & Jones, 1818), p. 3.

2. Genesis 2:7. Some scholars contend that this story derives directly from Near Eastern myths such as the Enuma Elish, a creation story that describes another god, Marduk, similarly creating man: "Blood to bone I form an original thing, its name is Man." See Barbara C. Sproul, *Primal Myths: Creation Myths Around the World* (San Francisco: Harper & Row, 1979), p. 104.

3. See the description in George Zarkadakis, *In Our Own Image* (New York: Pegasus Books, 2015), p. 33; and Apollonius Rhodius, *Argonautica*, trans. William H. Race (Cambridge, MA: Harvard University Press, 2008). A similar tale is told by Homer, and has been described by some as one of the earliest examples of artificial intelligence. See Harry Lewis, *Classics of Computer Science*, forthcoming.

4. See Justin Pollard and Howard Reid, *The Rise and Fall of Alexandria: Birthplace of the Modern World* (New York: Viking, 2006), p. 132. For Hero, see Noel Sharkey, "The Programmable Robot of Ancient Greece," *New Scientist*, July 7, 2007, pp. 32–35, at https://www.newscientist.com/issue/2611/; and A. G. Drachmann, "Hero's Windmill," *Centaurus* 7 (1961), pp. 145–51.

5. Joseph Needham, *Science and Civilization in China*: vol. 2, *History of Scientific Thought* (Cambridge, UK: Cambridge University Press, 1956), p. 53.

6. From Pindar's Seventh Olympic Ode. See Pindar, *Pindar*, trans. Rev. C. A. Wheelwright (New York: Harper & Brothers, 1864), p. 54.

7. For more on the development of automata during this era, see Silvio A. Bedini, "The Role of Automata in the History of Technology," *Technology and Culture*, vol. 5, no. 1 (Winter 1964), pp. 24–42; and Mark E. Rosheim, *Leonardo's Lost Robots* (Berlin: Springer, 2006), pp. 69–113. Another early creator was Al-Jazari, who worked across Anatolia (now Turkey) in the twelfth and thirteenth centuries. His works included a mechanical wine servant and a water-powered orchestra that could float on a lake and entertain party guests.

8. Indeed, the era of the early Industrial Revolution witnessed a widespread belief in "vitalism," the idea that humans were sustained by inner, invisible forces that were either linked to or similar to electricity. See the discussion in Zarkadakis, *In Our Own Image*, pp. 37–42.

9. Quoted in Thomas Jefferson Hogg, *Shelley at Oxford* (London: Methuen, 1904), p. 20.

10. Mary Shelley, "Introduction to the 1831 Edition," *Frankenstein* (London: Colburn and Bentley, 1831), p. x.

11. It is also what drives debates about how we treat other species: if animals act consciously, rather than simply instinctively, we tend to feel as if we should accord them a status that is closer to human. See, for example, Carl Safina, *Beyond Words: What Animals Think and Feel* (New York: Henry Holt, 2015).

12. For the history of its predecessors, see Martin Campbell-Henry, "Origin of Computing," *Scientific American*, vol. 301, no. 3 (September 2009), pp. 62–69.

13. See Simon Singh, *The Code Book: The Science of Secrecy from Ancient Egypt to Quantum Cryptography* (New York: Anchor Books, 1999), pp. 127–89; and Frank Carter, "The Turing Bombe," *The Rutherford Journal*, vol. 3 (2010), at http://www.rutherford journal.org/article030108.html.

14. For an excellent description of this process and the mechanics behind Turing's Bombe, see Singh, *The Code Book*, pp. 170–81.

15. For in-depth accounts of Turing and the effort at Bletchley Park, see Sinclair McKay, *The Secret Lives of Codebreakers: The Men and Women Who Cracked the Enigma Code at Bletchley Park* (New York: Plume, 2012); Michael Smith, *Station X: The Codebreakers of Bletchley Park* (London: Channel 4 Books, 1998); and Singh, *The Code Book*, pp. 143–89.

16. Turing's first reference to this test came in a 1950 paper in which he poses the question, "Can machines do what we (as thinking entities) can do?" See Alan Turing, "Computing Machinery and Intelligence," *Mind*, vol. 59, no. 236 (October 1950), pp. 433–66. For more on Turing and his life, see Andrew Hodges, *Alan Turing: The Enigma* (Princeton: Princeton University Press, 1983), and David Leavitt, *The Man Who Knew Too Much: Alan Turing and the Invention of the Computer* (New York: W. W. Norton, 2006). For more on the Turing test, see Robert Epstein and Grace Peters, eds., *The Turing Test Sourcebook: Philosophical and Methodological Issues in the Quest for the Thinking Computer* (Dordrecht and Boston: Kluwer, 2004).

17. For more on the robot's development and people's reaction to him, see Erico Guizzo, "How Aldebaran Robotics Built Its Friendly Humanoid Robot, Pepper," *IEEE Spectrum*, December 26, 2014, at http://spectrum.ieee.org/robotics/home-robots/how-aldebaran-robotics-built-its-friendly-humanoid-robot-pepper; and Agence France-Presse in Tokyo, "Pepper the Chatty Robot Makes Friends on First Day at Work," *Guardian*, June 6, 2014, at https://www.theguardian.com/technology/2014/jun/06/pepper-chatty-robot-japan-softbank.

18. Quoted in Shotaro Tani, "SoftBank's Son Has Big AI Ambitions," *Nikkei Asian Review*, June 22, 2016, at http://asia.nikkei.com/Business/Companies/Softbank-s-Son-has-big-AI-ambitions. Son has also been cited as believing that robots will outsmart humans within thirty years. See Sam Shead, "Softbank Billionaire Devoting 97% of 'Time and Brain' to AI," *Forbes*, August 6, 2018, at https://www.forbes.com/sites/samshead/2018/08/06/softbank-billionaire-devoting-97-of-time-and-brain-to-ai/#7b2181ab537d; and Catherine Clifford, "Billionaire CEO of SoftBank: Robots Will Have an IQ of 10,000 in 30 Years," October 25, 2017, at https://www.cnbc.com/2017/10/25/masayoshi-son-ceo-of-softbank-robots-will-have-an-iq-of-10000.html.

19. See, for example, Marvin Minsky, *The Emotion Machine* (New York: Simon & Schuster, 2006); Nick Bostrom, *Superintelligence: Paths, Dangers, Strategies* (Oxford: Oxford University Press, 2014); and Hans Moravec, *Mind Children: The Future of Robot and Human Intelligence* (Cambridge, MA: Harvard University Press, 1988).

20. Vernor Vinge, "The Coming Technological Singularity: How to Survive in the Post-Human Era," NASA VISION-21 Symposium, March 30–31, 1993. See also James Barrat, *Our Final Invention: Artificial Intelligence and the End of the Human Era* (New York: Thomas Dunne, 2013), pp. 196–97; and Yuval Noah Harari, *Sapiens: A Brief History of Humankind* (New York: Harper, 2015), pp. 397–414.

21. Quoted in Jeff Goodall, "The Rise of the Intelligent Machines: Part 1," *Rolling Stone*,

no. 1256 (March 10, 2016), p. 50. See also Samuel Gibbs, "Elon Musk Leads 116 Experts Calling for Outright Ban of Killer Robots," *Guardian*, August 20, 2017, at https://www.theguardian.com/technology/2017/aug/20/elon-musk-killer-robots -experts-outright-ban-lethal-autonomous-weapons-war; and Peter Holley, "Elon Musk's Nightmarish Warning: AI Could Become 'an Immortal Dictator from Which We Would Never Escape,'" *Washington Post*, April 6, 2018, at https://www .washingtonpost.com/news/innovations/wp/2018/04/06/elon-musks -nightmarish-warning-ai-could-become-an-immortal-dictator-from-which-we -would-never-escape/.

22. Musk later said that his remarks were overstated, but he reiterated his concern that "the rise of smart machines brings up serious questions that we need to consider about who we are as humans and what kind of future we are building for ourselves." Quoted in Goodall, "The Rise of the Intelligent Machines: Part 1," p. 48.

23. Quoted in ibid.

24. The old mantra ran: "Give a man a fish and he'll eat for a day. Teach a man to fish and he'll eat for a lifetime."

25. See Erik Brynjolfsson and Andrew McAfee, *The Second Machine Age: Work, Progress, and Prosperity in a Time of Brilliant Technologies* (New York: W. W. Norton, 2014), p. 32; and Martin Ford, *Rise of the Robots: Technology and the Threat of a Jobless Future* (New York: Basic Books, 2015), p. 16.

26. Andrew Dalton, "Foxconn Replaces 60,000 Human Workers with Robots," Engadget, at https://www.engadget.com/2016/05/25/foxconn-replaces-60000 -humans-workers-with-robots/.

27. Akshat Rathi, "WhatsApp Bought for $19 Billion, What Do Its Employees Get?" *The Conversation*, February 20, 2014, at http://theconversation.com/whatsapp -bought-for-19-billion-what-do-its-employees-get-23496.

28. Julie Sobowale, "How Artificial Intelligence Is Transforming the Legal Profession," *ABA Journal* (April 2016), p. 1.

29. See Huiying Liang et al., "Evaluation and Accurate Diagnoses of Pediatric Diseases Using Artificial Intelligence," *Nature Medicine*, vol. 25 (February 11, 2019), pp. 433–38, at https://www.nature.com/articles/s41591-018-0335-9.

30. Alyson Shontell, "A Startup That Uses Robots to Write News Gets Acquired for $80 Million in Cash," *Business Insider*, February 23, 2015, at http://www .businessinsider.com/automated-insights-gets-acquired-by-vista-for-80-million -2015-2.

31. The great conservative economist Friedrich Hayek first provided the theoretical backing for such a policy in the 1970s. See Hayek, *Law, Legislation, and Liberty*: vol. 3, *The Political Order of a Free People* (Chicago: University of Chicago Press, 1979), esp. pp. 54–55. For a contemporary model, see Jeffrey D. Sachs and Laurence J. Kotlikoff, "Smart Machines and Long-Term Misery," National Bureau of Economic Research, working paper no. 18629, December 2012.

32. See Rachel Minder, "Swiss Voters Reject Plan for Guaranteed Income," *New York Times*, June 6, 2016, p. A8; and Leonid Bershidsky, "Finland Basic Income Test Wasn't Ambitious Enough," *Bloomberg*, April 26, 2018, at https://www.bloomberg .com/opinion/articles/2018-04-26/finland-s-basic-income-experiment-was

-doomed-from-the-start. Switzerland had a referendum along these lines in 2016, which failed to pass.

33. He is also not alone. In 2017, over a million people asked Amazon's Alexa to marry them, and more than three thousand people have registered for commemorative marriage certificates with their favorite anime characters. Emiko Jozuka, "Beyond Dimensions: The Man Who Married a Hologram," CNN, December 29, 2018, at https://www.cnn.com/2018/12/28/health/rise-of-digisexuals-intl/index.html.

34. According to *The Economist*, Japan in 2014 already had a shortage of more than seven hundred thousand elder care workers. To address this gap, the government launched a program in 2013 that promises to pay two-thirds of the costs associated with developing inexpensive robots to help the elderly or their caregivers. See "Difference Engine: The Caring Robot," *Economist*, May 14, 2014, at http://www.economist.com/blogs/babbage/2013/05/automation-elderly.

35. Technically, roboticists refer to these kinds of robots as being "assistive" or "socially assistive." For a discussion, see Adriana Tapus, Maja J. Matarić, and Brian Scassellati, "The Grand Challenges in Socially Assistive Robotics," *IEEE Robotics and Automation Magazine*, vol. 14 (2007), pp. 1–7, at http://citeseerx.ist.psu.edu/viewdoc/download?doi=10.1.1.127.5625&rep=rep1&type=pdf; Roger Bemelmans, Gert Jan Gelderman, Pieter Jonker, and Luc de Witte, "Socially Assistive Robots in Elderly Care: A Systematic Review into Effects and Effectiveness," *Journal of the American Medical Directors Association*, vol. 13, no. 2 (February 2012), pp. 114–20; and D. Feil-Seifer and Maja J. Matarić, "Defining Socially Assistive Robots," *9th International Conference on Rehabilitation Robotics*, 2005, pp. 465–68.

36. See J. Mark Lytle, "Honda Creates 3D CPU to Power Super Robots," *Techradar*, January 30, 2008, at http://www.techradar.com/news/computing-components/processors/honda-creates-3d-cpu-to-power-super-robots-215259; and Lee Ann Obringer and Jonathan Strickland, "Honda ASIMO Robot," *How Stuff Works*, April 11, 2007, at http://science.howstuffworks.com/asimo1.htm.

37. See "Developing KASPAR," University of Hertfordshire, at http://www.herts.ac.uk/kaspar/introducing-kaspar; "Robots in the Classroom Help Autistic Children Learn," BBC, November 8, 2012, available at http://www.bbc.co.uk/new/edicatopm-20252593; and Joshua Wainer, Ben Robbins, Farshid Amirabdollahian, and Kerstin Dautehahn, "Using Humanoid Robot KASPAR to Autonomously Play Triadic Games and Facilitate Collaborative Play Among Children with Autism," *IEEE Transactions on Autonomous Mental Development*, vol. 6, no. 3 (September 2014), pp. 183–99.

38. Adam Cohen, "The Machine Nurturer," *Time*, December 3, 2000, pp. 110–13.

39. Ibid.; and "Kismet," Robotics Today, at http://www.roboticstoday.com/robots/kismet. More generally, see Cynthia Breazeal, *Designing Sociable Robots* (Cambridge, MA: MIT Press, 2002).

40. See Cynthia Breazeal and Brian Scassellati, "Infant-like Social Interactions Between a Robot and a Human Caretaker," *Adaptive Behavior*, vol. 8, no. 1 (January 2000), pp. 49–74.

41. J. Wada et al., "Analysis of Factors That Bring Mental Effects to Elderly People in

Robot Assisted Activity," *Proceedings of IEEE/RSJ International Conference on Intelligent Robots and Systems (IROS '02)*, Lausanne, September 2000, pp. 1152–57.

42. C. G. Burgar et al., "Development of Robots for Rehabilitation Theory: The Palo Alto vs. Stanford Experience," *Journal of Rehabilitation Research and Development*, vol. 37, no. 6 (2000), pp. 639–52; J. Eriksson, M. J. Matarić, and C. Winstein, "Hands-Off Assistive Robotics for Post-Stroke Arm Rehabilitation," *Proceedings of IEEE International Conference on Rehabilitation Robotics*, Chicago, June 2005, pp. 21–24; and K. I. Kang, M. J. Matarić, Matk J. Cunningham, and Becky Lopez, "A Hands-Off Physical Therapy Assistance Robot for Cardiac Patients," *Proceedings of IEEE International Conference on Rehabilitation Robotics*, Chicago, June 2005, pp. 337–40.

43. See, for example, I. P. Werry and K. Dautenhahn, "Applying Mobile Robot Technology to the Rehabilitation of Autistic Children," *Proceedings of the 7th International Symposium on Intelligent Systems*, Coimbra, Portugal, July 1999, pp. 265–22; F. Michaud and A. Clavet, "Robotoy Contest—Designing Mobile Robotic Toys for Autistic Children," *Proceedings of the American Society for Engineering Education*, Albuquerque, 2001; and Brian Scassellati, Henny Admoni, and Maja Matarić, "Robots for Use in Autism Research," *Annual Review of Biomedical Engineering*, vol. 14 (August 2012), pp. 275–94.

44. See Nellie Bowles, "Human Contact Is Now a Luxury Good," *New York Times*, May 23, 2019, at https://www.nytimes.com/2019/03/23/sunday-review/human-contact-luxury-screens.html.

45. See John H. Richardson, "How Much Better Can We Stand to Be?" *Esquire*, May 2015, pp. 89–95, 130. Viv was subsequently purchased by Samsung, and the underlying technology was integrated into its personal assistant software. For more recent developments with personal assistant technologies, see Megan Wollerton, "Alexa, Google Assistant and Siri Will Get Smarter This Year," *Cnet*, June 7, 2019, at https://www.cnet.com/news/alexa-vs-google-assistant-vs-siri-the-state-of-voice-after-google-io-and-wwdc-2019/.

46. John Markoff and Paul Mazur, "For Sympathetic Ear, More Chinese Turn to Smartphone Program," *New York Times*, July 31, 2015, at http://www.nytimes.com/2015/08/04/science/for-sympathetic-ear-more-chinese-turn-to-smartphone-program.html?_r=0.

47. Yongdong Wang, "Your Next New Best Friend Might Be a Robot," *Nautilus*, February 4, 2016, at http://nautil.us/issue/33/attraction/your-next-new-best-friend-might-be-a-robot.

48. See Ma Si, "Microsoft Expands Presence of AI Platform Xiaoice," *China Daily*, July 28, 2018, at http://www.chinadaily.com.cn/a/201807/28/WS5b5baf5ea31031a351e90b14.html.

49. See Ellie Zolfagharifard and Julian Robinson, "The 'World's Sexiest Robot' Revealed," *Daily Mail*, November 27, 2015, at http://www.dailymail.co.uk/sciencetech/article-3335552/The-world-s-sexiest-robot-revealed-Eerily-life-like-female-android-turns-heads-China.html.

50. Maria Khan, "Japan: Humans Might Soon Be Taking Life-Like Androids as Partners in Marriage," at http://www.ibtimes.co.uk/japan-humans-might-soon-be-taking-life-life-androids-partners-marriage-1475838.

51. Lovotics, at https://sites.google.com/site/lovoticsrobot/.

52. "Sex Dolls That Talk Back," *New York Times*, June 11, 2015, at http://www.nytimes.come/2015/06/12/technology/robotica-sex-robot-realdoll.html.

53. Interestingly, the campaign against sex robots has already been waged. See Campaign Against Sex Robots, at http://www.campaignagainstsexrobots.com/; and John Danaher, Brian D. Earp, and Anders Sandberg, "Should We Campaign Against Sex Robots?" in *Robot Sex: Social and Ethical Implications*, ed. J. Danaher and N. McArthur (Cambridge, MA: MIT Press, 2017).

54. Erik Sofge, "The Uncertain Future for Social Robots," *Popular Mechanics*, January 27, 2010, at http://www.popularmechanics.com/technology/robots/a5037/4343892/.

55. One of the earliest and most widespread instances of this phenomenon occurred with the Tamagotchi, an egg-shaped toy popular in the 1990s. Even though it only relayed its needs to the child using it, and otherwise didn't do much at all, children were famously attached to their Tamagotchis and eager to keep them "alive." See James Vlahos, "Artificially Yours," *New York Times Magazine*, September 20, 2015, pp. 44–49, 75–76; and Sofge, "The Uncertain Future for Social Robots." See also Sherry Turkle, "Authenticity in the Age of Digital Companions," *Interaction Studies*, vol. 8, no. 3 (2007), pp. 501–17.

56. In 1992, Francis Fukuyama famously prophesied, in his book of the same name, the "end of history." He was wrong.

9. Engineering the End of Death

1. *Thus Spake Zarathustra*, Part III ("The Convalescent"), pp. 329–30, as cited in Herbert Marcuse, *Eros and Civilization: A Philosophical Inquiry into Freud*, 2nd ed. (Boston: Beacon Press, 1966), p. 122.

2. Vernor Vinge, "The Coming Technological Singularity: How to Survive in the Post-Human Era," *Vision-21: Interdisciplinary Science and Engineering in the Era of Cyberspace*, NASA Conference Publication 10129, NASA Lewis Research Center (1993), pp. 11–22.

3. Sigmund Freud, *Beyond the Pleasure Principle* (New York: W. W. Norton, 1990), p. 46.

4. These figures, both historical and contemporary, are heavily affected by rates of infant and child mortality. Once an individual reaches adulthood, their expected life span is considerably higher than that of the overall population. See H. Beltrán-Sánchez, E. M. Crimmins, and C. E. Finch, "Early Cohort Mortality Predicts the Rate of Aging in the Cohort: A Historical Analysis," *Journal of Developmental Origins of Health and Disease*, vol. 3, no. 5 (October 2012), pp. 380–86.

5. See Dan Buettner, *The Blue Zones: Lessons for Living Longer from the People Who've Lived the Longest* (Washington, DC: National Geographic Books, 2008).

6. Debt is higher because older populations are demanding (and receiving) greater spending on social services. Growth is lower because the country's workforce is shrinking. See Gregg Easterbrook, "What Happens When We All Live to 100?" *Atlantic*, October 2014, at https://www.theatlantic.com/magazine/archive/2014/10/what-happens-when-we-all-live-to-100/379338/.

7. Some scientists have hypothesized that humans have a fixed life span of about 115 years. Others, however, are not so sure. See Xiao Dong, Brandon Milholland,

and Jan Vijg, "Evidence for a Limit to Human Lifespan," *Nature*, vol. 538 (October 5, 2016), pp. 257–59; and Carl Zimmer, "How Long Can We Live?" *New York Times*, June 28, 2018, at https://www.nytimes.com/2018/06/28/science/human-age-limit.html.

8. See for example, "Prosthetic Limbs, Controlled by Thought," *New York Times*, May 20, 2015, at http://www.nytimes.com/2015/05/21/technology/a-bionic-approach -to-prosthetics-controlled-by-thought.html.

9. Violence, accidents, and congenital conditions kill people in the first decades of life. Cancer, heart disease, and stroke come later. See Melonie Heron, "Deaths: Leading Causes for 2016," *National Vital Statistics Reports*, vol. 67, no. 6 (Hyatts-ville, MD: National Center for Health Statistics, 2018).

10. For a review of various theories and their connections, see Thomas B. L. Kirk-wood, "Understanding the Odd Science of Aging," *Cell*, vol. 120, no. 4 (February 25, 2005), pp. 437–47. For a full and fascinating scientific description, see Ronald S. Petralia et al., "Aging and Longevity in the Simplest Animals and the Quest for Immortality," *Ageing Research Review*, vol. 16 (July 2014), pp. 66–82.

11. See, for example, R. E. Ricklefs, "Evolutionary Theories of Aging: Confirmation of a Fundamental Prediction, with Implications for the Genetic Basis and Evolu-tion of Life Span," *The American Naturalist*, vol. 152, no. 1 (July 1998), pp. 24–44.

12. Malignant cells can divide indefinitely. For more on the link between cancer re-sistance and aging, see Judith Campisi, "Aging, Tumor Suppression and Cancer: High Wire-Act!" *Mechanisms of Ageing and Development*, vol. 126, issue 1 (January 2005), pp. 51–58.

13. For the role stress plays in shrinking telomeres, see Thomas von Zglinicki, "Ox-idative Stress Shortens Telomeres," *Trends in Biochemical Science*, vol. 27, no. 7 (July 2002), pp. 339–44.

14. This process, known as pruning, helps explain why children can often "lose" a language they once spoke frequently. If they don't use that language as they enter puberty, the neural connections that supported it can physically disap-pear. See Peter R. Huttenlocher, *Neural Plasticity* (Cambridge, MA: Harvard University Press, 2002), p. 207.

15. See Ruth Peters, "Ageing and the Brain," *Postgraduate Medical Journal*, vol. 82, no. 964 (February 2006), pp. 84–88; Lars Svennerholm, Kerstin Boström, and Birgitta Jungbjer, "Changes in Weight and Composition of Major Membrane Compo-nents of Human Brain During the Span of Adult Human Life of Swedes," *Acta Neuropathologica*, vol. 94, no. 4 (September 1997), pp. 345–52; N. Raz, "The Age-ing Brain: Structural Changes and Their Implications for Cognitive Ageing," in *New Frontiers in Cognitive Ageing*, ed. R. Dixon et al. (Oxford: Oxford University Press, 2004); and Michael A. Lodato et al., "Aging and Neurodegeneration Are Associated with Increased Mutations in Single Human Neurons," *Science*, vol. 359, no. 6375 (February 2, 2018), pp. 555–59.

16. For an overview, see "Inside the Brain: A Tour of How the Mind Works," Alzheimer's Association, at https://www.alz.org/alzheimers-dementia/what-is-alzheimers /brain_tour. For greater scientific detail, see Daniel P. Perl, "Neuropathology of Alzheimer's Disease," *Mount Sinai Journal of Medicine*, vol. 77, no. 1 (January/ February 2010), pp. 32–42; and J. Pozueta, R. Lefort, and M. L. Shelanski, "Synaptic

Changes in Alzheimer's Disease and Its Models," *Neuroscience*, vol. 251 (October 22, 2013), pp. 51–65.

17. A. Lobo et al., "Prevalence of Dementia and Major Subtypes in Europe: A Collaborative Study of Population-Based Cohorts," *Neurology*, vol. 54, supplement 5 (June 2000), pp. S4–S9; and Aharon W. Zorea, *Finding the Fountain of Youth: The Science and Controversy Behind Extending Life and Cheating Death* (Santa Barbara: Greenwood, 2017), p. 215.

18. See Elizabeth Blackburn and Elissa Epel, *The Telomere Effect: A Revolutionary Approach to Living Younger, Healthier, Longer* (New York: Grand Central Publishing, 2017, p. xv.

19. See Jamie Metzl, *Hacking Darwin: Genetic Engineering and the Future of Humanity* (New York: Sourcebooks, 2019), p. 139.

20. David Ferry, *Gilgamesh: A New Rendering in English Verse* (New York: Farrar, Straus, and Giroux, 1992).

21. Quoted in Samuel Eliot Morison, *The European Discovery of America: The Southern Voyages 1492–1616* (New York: Oxford University Press, 1974), p. 504. Subsequent historians have argued that Ponce de León himself was never really in search of the fountain, but the quest was attached to him by early chroniclers of the Spanish expeditions and has remained there ever since. See Zorea, *Finding the Fountain of Youth*, pp. 11–38; Michael J. Francis, "Who Started the Myth About a Fountain of Youth?" *Forum: The Magazine of the Florida Humanities Council*, vol. 35, no. 3 (Fall 2011), pp. 6–9; and Douglas T. Peck, "Anatomy of an Historical Fantasy: The Ponce de León–Fountain of Youth Legend," *Revista de Historia de América*, no. 123 (January–December 1998), pp. 63–87.

22. More precisely, Moore described a doubling in the number of transistors in a dense integrated circuit. See Gordon E. Moore, "Cramming More Components onto Integrated Circuits," *Electronics*, vol. 38, no. 8 (April 19, 1965), pp. 114–17.

23. For a basic scientific overview of telomeres and telomerase, see Elizabeth Blackburn et al., "Human Telomere Biology: A Contributory and Interactive Factor in Aging, Disease Risks, and Protection," *Science*, vol. 350, no. 6265 (December 4, 2015), pp. 1193–98; and E. H. Blackburn and J. W. Szostak, "The Molecular Structure of Centromeres and Telomeres," *Annual Review of Biochemistry*, vol. 53 (1984), pp. 163–94. For a more popular treatment, see Blackburn and Epel, *The Telomere Effect*. Dr. Blackburn won the Nobel Prize in Medicine in 2009 for her discovery of telomeres and their effects.

24. Bernard Strehler, quoted in Zorea, *Finding the Fountain of Youth*, p. 109.

25. See Nathaniel Rich, "Forever and Ever," *New York Times Magazine*, December 2, 2012, pp. 33–39, 66, 68, 70, 78; and S. Piraino et al., "Reversing the Life Cycle: Medusae Transforming into Polyps and Cell Transdifferentiation in *Turritopsis nutricula* (Cnidaria, Hydrozoa)," *The Biological Bulletin*, vol. 190, no. 3 (June 1996), pp. 302–12.

26. N. Funayama, "The Stem Cell System in Demosponges: Suggested Involvement of Two Types of Cells," *Development Genes and Evolution*, vol. 223, no. 1–2 (March 2013), pp. 23–38; and Ronald S. Petralia et al., "Aging and Longevity in the Sim-

plest Animals and the Quest for Immortality," *Ageing Research Review*, vol. 16 (July 2014), pp. 66–82.

27. Cynthia Kenyon, "The Plasticity of Aging: Insights from Long-Lived Mutants," *Cell*, vol. 120, no. 4 (February 25, 2005), pp. 449–60. The original research is in Cynthia Kenyon et al., "A *C. Elegans* Mutant That Lives Twice as Long as Wild Type," *Nature*, vol. 366, no. 6454 (December 2, 1993), pp. 461–64; and Nuno Arantes-Oliveira et al., "Healthy Animals with Extreme Longevity," *Science*, vol. 302, no. 5645 (October 24, 2003), p. 611.

28. Quoted in Ian Sample, "Harvard Scientists Reverse the Ageing Process in Mice—Now for Humans," *Guardian*, November 28, 2010, at https://www.theguardian .com/science/2010/nov/28/scientists-reverse-ageing-mice-humans; and Richard Saltus, "Partial Reversal of Aging Achieved in Mice," *Harvard Gazette*, November 28, 2010, at https://harvard.edu/gazette.story/2010/11/partial-reversal-of-aging -achieved-in-mice/. For the full study, see M. Jaskelioff et al., "Telomerase Reactivation Reverses Tissue Degeneration in Aged Telomerase-Deficient Mice," *Nature*, vol. 469, no. 7328 (January 6, 2011), pp. 102–106.

29. See A. G. Bodnar et al., "Extension of Life-Span by Introduction of Telomerase into Normal Human Cells," *Science*, vol. 279, no. 5349 (January 16, 1998), pp. 349–52. For a broader analysis of how humans might change their behavior in ways that would also help their telomeres, see Blackburn and Epel, *The Telomere Effect*.

30. For a scientific review, see Calvin Harley et al., "A Natural Product Telomerase Activator as Part of a Health Maintenance Program," *Rejuvenation Research*, vol. 14, no. 1 (February 2011), pp. 45–56.

31. Quoted in Sample, "Harvard Scientists Reverse the Ageing Process in Mice—Now for Humans."

32. See "MIT Researchers Uncover New Information About Anti-Aging Gene," *MIT News*, February 16, 2000, at http://news.mit.edu/2000/guarente; and Nicholas Wade, "Longer Lives for Obese Mice, with Hope for Humans of All Sizes," *New York Times*, August 18, 2011, at https://www.nytimes.com/2011/08/19/science/19fat.html.

33. For a scientific review, see A. Salminen and K. Kaarnirante (2012), "AMP-Activated Protein Kinase (AMPK) Controls the Aging Process via an Integrated Signaling Network," *Ageing Research Reviews*, vol. 11, no. 2 (April 2012), pp. 230–41.

34. Luigi Fontana, Linda Partridge, and Valter Longo, "Dietary Restriction, Growth Factors and Aging: From Yeast to Humans," *Science*, vol. 328, no. 5976 (April 16, 2010), pp. 321–26.

35. Quoted in Tad Friend, "Silicon Valley's Quest to Live Forever," *New Yorker*, April 3, 2017, p. 54.

36. Aubrey de Grey with Michael Rae, *Ending Aging: The Rejuvenation Breakthroughs That Could Reverse Human Aging in Our Lifetime* (New York: St. Martin's Press, 2008), p. 21.

37. Ibid., p. 44.

38. Methuselah Foundation, at https://www.mfoundation.org/. For more on de Grey and his theories, see de Grey with Rae, *Ending Aging*; and Sherwin Nuland, "Live Forever? Aubrey de Grey Thinks He Can Defeat Death. Is He Nuts?" *MIT Technology*

Review, February 1, 2005, at https://www.technologyreview.com/s/403654/do-you
-want-to-live-forever/.

39. René Descartes, *Discourse on Method*, trans. F. Sutcliffe (London: Penguin Books, 2005), p. 54.

40. Plato, *Dialogues of Plato*, trans. Benjamin Jowett (New York: D. Appleton, 1898), p. 424.

41. Thomas Aquinas, *On Human Nature* (Indianapolis: Hackett Publishing, 1999), p. 64.

42. This is the part, Descartes wrote, "whose whole essence or nature is simply to think." Descartes, *Discourse on Method*, p. 54.

43. Martine had one son already from a brief Kenyan relationship; Bina (then called Beverlee) had a daughter. They adopted each other's children and had two more together.

44. Quoted in Lisa Miller, "The Trans-Everything CEO," *New York*, September 7, 2014, at http://nymag.com/news/features/martine-rothblatt-transgender-ceo/index1 .html#print.

45. Ibid.

46. Conversation with author, September 20, 2018, Silver Spring, MD.

47. Lifenaut is not the only website offering digital immortality. Indeed, several firms have jumped into this space, offering users various ways of uploading material in life that can be stored, shared, and perhaps even reanimated after their death. See Laura Parker, "How to Become Virtually Immortal," *New Yorker*, April 4, 2014, at http://www.newyorker.com/tech/elements/how-to-become-virtually -immortal; and Chris Gayomali, "Eterni.me Wants to Let You Skype Your Family After You're Dead," *Fast Company*, January 30, 2014, at https://www .fastcompany.com/3025797/eternime-wants-to-let-you-skype-your-family-from -the-grave. On the more general issues surrounding digital storage of memories, see Alec Wilkinson, "Remember This?" *New Yorker*, May 28, 2007, at http:// newyorker.com/magazine/2007/05/28/remember-this.

48. See Jessica Roy, "The Rapture of the Nerds," *Time*, April 17, 2014, at http://time .com/66536/terasem-transcendence-religion-technology/.

49. See Martine Rothblatt, "The Terasem Mind Uploading Experiment," *International Journal of Machine Consciousness*, vol. 4, no. 1 (2012), p. 142.

50. Other key contributors to the ideas behind machine consciousness and transhumanism include Marvin Minsky, Nick Bostrom, John von Neumann, and Hans Moravec. Many of the guiding principles come from the earlier work of Alan Turing and the science fiction writer Isaac Asimov. See Marvin Minsky, *The Emotion Machine* (New York: Simon & Schuster, 2006); Nick Bostrom, *Superintelligence: Paths, Dangers, Strategies* (Oxford: Oxford University Press, 2014); John von Neumann, *The Computer and the Brain* (New Haven: Yale University Press, 1958); Hans Moravec, *Mind Children: The Future of Robot and Human Intelligence* (Cambridge, MA: Harvard University Press, 1988); Alan Turing, "Computing Machinery and Intelligence," *Mind*, vol. 59, no. 236 (October 1950), pp. 434–60; and Isaac Asimov, *I, Robot* (New York: Doubleday, 1950).

51. Ray Kurzweil, *The Age of Spiritual Machines* (New York: Penguin Books, 1999), p. 22.

52. Ibid., pp. 101–56. See also Ray Kurzweil, *How to Create a Mind: The Secret of Human Thought Revealed* (New York: Penguin Books, 2012).

53. Kurzweil, *The Age of Spiritual Machines*, p. 153.

54. Ray Kurzweil, *The Singularity Is Near* (New York: Penguin Books, 2006), pp. 40–41. Italics in the original.

55. See, for example, John Searle, "I Married a Computer," *New York Review of Books*, April 8, 1999, pp. 34–38; Colin McGinn, "Hello HAL," *New York Times*, January 3, 1999, pp. 11–12; and Paul Allen, "The Singularity Isn't Near," *MIT Technology Review*, October 12, 2011, at https://www.technologyreview.com/s/425733/paul-allen-the-singularity-isnt-near/.

56. Kurzweil and his followers have also devoted a great deal of time, energy, and research to the evolving potential of nanobots, a topic not covered here. See Ray Kurzweil and Terry Grossman, *Fantastic Voyage: Live Long Enough to Live Forever* (Plume, 2005).

57. Jon Cohen, "Memory Implants: A Maverick Scientist Believes He Has Deciphered the Code by Which the Brain Forms Long-Term Memories," *MIT Technology Review*, at https://www.technologyreview.com/s/513681/memory-implants/; and Theodore W. Berger et al., "Brain-Implantable Biomimetic Electronics as the Next Era in Neural Prosthetics," *Proceedings of the IEEE*, vol. 89, no. 7 (July 2001), pp. 993–1010. For similar efforts by scientists at IBM to build a computer "inspired by the brain," see Dharmendra Modha, "Introducing a Brain-Inspired Computer," at http://www.research.ibm.com/articles/brain-chip.shtml.

58. Kenneth J. Hayworth, "Electron Imaging Technology for Whole Brain Neural Circuit Mapping," *International Journal of Machine Consciousness*, vol. 4, no. 1 (2012), p. 89. For more general but similar predictions, see Ray Kurzweil, "The Coming Merging of Mind and Machine," *Scientific American*, March 23, 2009, at https://www.scientificamerican.com/article/merging-of-mind-and-machine/; for related efforts to map one region of the hypothalamus, see Jeffrey R. Moffitt et al., "Molecular, Spatial, and Functional Single-Cell Profiling of the Hypothalamic Preoptic Region," *Science*, vol. 362, no. 6416 (November 16, 2018), pp. 1–21.

59. See National Institutes of Health and Department of Energy, "Understanding Our Genetic Inheritance. The US Human Genome Project: The First Five Years, FY 1991–1995" (Washington, DC: National Institutes of Health, U.S. Department of Health and Human Services, and U.S. Department of Energy, 1990); and Francis Collins and David Gallas, "A New Five-Year Plan for the U.S. Human Genome Project," *Science*, vol. 262, no. 5130 (October 1, 1993), pp. 43–46. Kenneth Hayworth, author of the study cited in the previous note, also cites the U.S. space program of the early 1960s as an example of a breakthrough technology that becomes possible once sufficient power—both literal and figurative—is put behind it.

60. Erika Check, "James Watson's Genome Sequenced," *Nature*, June 1, 2007, at https://www.nature.com/news/2007/070528/full/news070528-10.html; and Chelsea Gohd, "Soon, You Could Have Your Genome Sequenced in 60 Minutes for $100," Futurism.com, January 11, 2017, at https://futurism.com/soon-you-could-have-your-genome-sequenced-in-60-minutes-for-100.

61. See, for example, Vernor Vinge, "Signs of the Singularity," *IEEE Spectrum*, vol. 45, no. 6 (2008), pp. 76–82; and John Storrs Hall, "Self-Improving AI: An Analysis," *Minds and Machines*, vol. 17, no. 3 (October 2007), pp. 249–59.

62. Several projects along these lines are already underway, including the European Union's Human Brain Project, which aims to build a complete computer model of a functioning brain. As George Zarkadakis notes, the project is headquartered along the banks of Lake Geneva, just miles from the villa in which Mary Shelley wrote *Frankenstein*. See George Zarkadakis, *In Our Own Image* (New York: Pegasus Books, 2015), pp. 164–65.

63. Kurzweil, "The Coming Merging of Mind and Machine."

64. As of 2018, computer scientists at Terasem are using chatbot software to develop operating systems that they hope will eventually replicate the human consciousness behind each user's mindfiles. Rothblatt, "The Terasem Mind Uploading Experiment," p. 147.

65. See Ashlee Vance, "Merely Human? So Yesterday," *New York Times*, June 13, 2010, pp. BU1, BU6–7.

66. Interview with author, September 20, 2018, Silver Spring, MD.

67. Kurzweil, *The Singularity Is Near*.

68. Rothblatt, "The Terasem Mind Uploading Experiment," p. 141.

69. Robert Browning, "A Toccata of Galuppi's," *Men and Women* (New York: H. M. Caldwell, 1902), p. 57.

70. Tad Friend, "Silicon Valley's Quest to Live Forever," p. 65; and Maureen Dowd, "Elon Musk's Future Shock," *Vanity Fair*, April 2017, p. 116. For a broader review, see S. Jay Olshansky and Bruce A. Carnes, *The Quest for Immortality: Science at the Frontiers of Aging* (New York: W. W. Norton, 2001).

71. Eric Neumayer and Thomas Plümper, "Inequalities of Income and Inequalities of Longevity: A Cross-Country Study," *American Journal of Public Health*, vol. 106, no. 1 (January 2016), pp. 160–65.

72. One shudders to think about the argument to define mortality as a "pre-existing condition."

73. See Easterbrook, "What Happens When We All Live to 100?" Here Easterbook is quoting Sheila Smith, a Japan specialist at the Council on Foreign Relations. For data, see "Life Expectancy at Birth, Total (Years)—Bangladesh, Japan," World Bank, at https://data.worldbank.org/indicator/SP.DYN.LE00.IN?locations=BD-JP.

74. Rothblatt, "The Terasem Mind Uploading Experiment," p. 143.

75. Sigmund Freud, *Reflections on War and Death* (New York: Moffat, Yard, 1918), p. 41.

Conclusions: Welcome to Tomorrowland

1. Shulamith Firestone, *The Dialectic of Sex* (New York: Farrar, Straus and Giroux, 1970), pp. 175, 184.

2. See, for instance, Erwin Loh, "Medicine and the Rise of the Robots: A Qualitative Review of Recent Advances of Artificial Intelligence in Health," *BMJ Leader*, vol. 2, no. 2 (June 2018), at https://bmjleader.bmj.com/content/leader/2/2/59.full.pdf.

3. Dana Varinsky, "These Robots Are Milking Cows Without Any Humans Involved, and the Cows Seem Into It," *Business Insider,* June 3, 2017 at https://www.businessinsider.com/automation-dairy-farms-robots-milking-cows-2017-6.

4. Marc Bain, "A New T-Shirt-Sewing Robot Can Make as Many Shirts per Hour as 17 Factory Workers," *Quartz,* August 30, 2017, at https://qz.com/1064679/a-new-t-shirt-sewing-robot-can-make-as-many-shirts-per-hour-as-17-factory-workers/.

5. As *New York Times* columnist Farhad Manjoo wrote in 2015: "You may not be contemplating becoming an Uber driver any time soon, but the Uberization of work may soon be coming to your chosen profession." Farhad Manjoo, "Uber's Business Model Could Change Your Work," *New York Times,* January 28, 2015, at https://www.nytimes.com/2015/01/29/technology/personaltech/uber-a-rising-business-model.html.

6. For more on the general implications of the gig economy, see Sarah Kessler, *Gigged: The End of the Job and the Future of Work* (New York: St. Martin's Press, 2018); Jeremias Prassl, *Humans as a Service: The Promise and Perils of Work in the Gig Economy* (Oxford: Oxford University Press, 2018); and McKinsey Global Institute, *What The Future of Work Will Mean for Jobs, Skills, and Wages,* November 2017, at https://www.mckinsey.com/global-themes/future-of-organizations-and-work/what-the-future-of-work-will-mean-for-jobs-skills-and-wages.

7. Between 1983 and 2013, the percentage of Americans covered by a company's defined-benefit pension plans fell from 62 percent to 17 percent. See Center for Retirement Research at Boston College, Frequently Requested Data, "Workers with Pension Coverage by Type of Plan, 1983, 1992, 2001, and 2013," at http://crr.bc.edu/wp-content/uploads/1012/01/figure-15.pdf. More generally, see also Jacob Hacker, *The Great Risk Shift: The New Economic Insecurity and the Decline of the American Dream* (Oxford: Oxford University Press, 2008).

8. At the elite end of the labor market, this means that marketing managers at Facebook can take as many vacation days as they desire—so long as they're available whenever they're needed. At places like Starbucks or the Apple store, it means that workers often have their hours assigned and adjusted at the very last minute. See, for example, Jodi Kantor, "Working Anything but 9 to 5," *New York Times,* August 13, 2014, at https://www.nytimes.com/interactive/2014/08/13/us/starbucks-workers-scheduling-hours.html.

9. Usually defined as individuals born between 1980 and 1995.

10. Firestone, *The Dialectic of Sex,* p. 180. Italics in the original.

11. Conversation with author, Boston, MA, December 10, 2018.

12. For a rather melancholy exploration of some of these options, see Emily Witt, *Future Sex* (New York: Farrar, Straus and Giroux, 2016).

13. Firestone, *The Dialectic of Sex,* p. 187.

14. "Crude Marriage Rate, Selected Years, 1960–2016," Eurostat: Statistics Explained, June 27, 2018, at https://ec.europa.edu/eurostat/statistics-explained/index.php?title=File:Crude_marriage_rate,_selected_years,_1960-2016_(per_1_000 persons).png.

15. Judith Shulevitz, "Why Is Dating in the App Era Such Hard Work?" *Atlantic,* vol. 318, no. 4 (November 2016), pp. 52–54.

16. Data from Japan Ministry of Internal Affairs and Communications Statistics Bureau, *Japan Statistical Yearbook 2019*, chapter 2: "Populations and Households," table 2–16: "Live Births, Death, Foetal Deaths, Marriages and Divorces (1925 to 2016)," p. 71.

17. Josué Ortega and Philipp Hergovich, "The Strength of Absent Ties: Social Integration via Online Dating," September 14, 2018, at https://arxiv.org/pdf/1709.10478 .pdf; "First Evidence That Online Dating Is Changing the Nature of Society," *MIT Technology Review*, October 10, 2017, at https://www.technologyreview.com/s /609091/first-evidence-that-online-dating-is-changing-the-nature-of-society/; "Interracial Marriage: Who Is Marrying Out?" Pew Research Center, June 12, 2015, at http://www.pewresearch.org/fact-tank/2015/06/12/interracial-marriage-who-is -marrying-out/; and "What Does the 2011 Census Tell Us About Inter-Ethnic Relationships?" UK Office for National Statistics, July 3, 2014.

18. See Ean Higgins, "Three in Marriage Bed More of a Good Thing," *Australian*, December 10, 2011, at http://www.theaustralian.com.au/news/features/three -in-marriage-bed-more-of-a-good-thing/story-e6frg6z6-1226218569577.

19. See Debora L. Spar, "The Egg Trade: Making Sense of the Market for Human Oocytes," *New England Journal of Medicine*, vol. 356, no. 13 (March 2007), pp. 1289–91.

20. Interview with author, Cambridge, MA, November 29, 2016.

21. Interview with author, Oxford, England, January 28, 2016.

22. See Catherine Clifford, "Elon Musk: 'Mark My Words—A.I. Is Far More Dangerous Than Nukes,'" CNBC, March 13, 2018, at https://www.cnbc.com/2018/03/13 /elon-musk-at-sxsw-a-i-is-more-dangerous-than-nuclear-weapons.html; and Kevin Rawlinson, "Microsoft's Bill Gates Insists AI Is a Threat," BBC, January 29, 2015, at https://www.bbc.com/news/31047780.

23. See David Baltimore et al., "A Prudent Path Forward for Genomic Engineering and Germline Gene Modification," *Science*, vol. 348, issue 6230 (April 3, 2015), pp. 36–38; and Jennifer A. Doudna and Samuel H. Sternberg, *A Crack in Creation* (Boston: Houghton Mifflin Harcourt, 2017), esp. p. 188.

24. See, for example, Thomas C. Schelling, "A World Without Nuclear Weapons," *Daedalus*, vol. 138, no. 4 (Fall 2009), pp. 124–29; and Scott D. Sagan, "Why Do States Build Nuclear Weapons? Three Models in Search of a Bomb," *International Security*, vol. 21, no. 3 (Winter 1996–97), pp. 54–86.

25. See Michael Grothaus, "Bet You Didn't See This Coming: 10 Jobs That Will Be Replaced by Robots," *Fast Company*, January 9, 2017, at https://www.fastcompany .com/30667279/you-didnt-see-this-coming-10-jobs-that-will-be-replaced-by -robots.

26. Sebastian Mohr and Lene Koch, "Transforming Social Contracts: The Social and Cultural History of IVF in Denmark," *Reproductive Biomedicine & Society Online*, vol. 2 (June 2016), pp. 88–96.

27. Daphna Birenbaum-Carmeli, "Thirty-Five Years of Assisted Reproductive Technologies in Israel," *Reproductive Biomedicine & Society Online*, vol. 2 (June 2016), pp. 16–23.

28. This was the total number of options listed by Facebook as of 2017. See Paul Ken-

gor, "71 Gender Options: Oh, Boy!," *TribLive*, April 1, 2017, at https://triblive.com/opinion/featuredcommentary/12128195-74/71-gender-options-oh-boy.

29. Yet, interestingly, the British government recently created a Minister for Loneliness. See Rebecca Mead, "What Britain's 'Minister of Loneliness' Says About Brexit and the Legacy of Jo Cox," *New Yorker*, January 28, 2018, at https://www.newyorker.com/culture/cultural-comment/britain-minister-of-loneliness-brexit-jo-cox.

30. For two recent proposals in support of UBI from very different ideological positions, see Charles Murray, *In Our Hands: A Plan to Replace the Welfare State* (Washington, DC: The AEI Press, 2006); and Andy Stern, *Raising the Floor: How Universal Basic Income Can Renew Our Economy and Rebuild the American Dream* (New York: Public Affairs, 2016).

31. Mary Shelley, *Frankenstein* (1818; reprint, New York: Pocket Books, 2004), p. 244.

Acknowledgments

This book unfolded over many years, written between a series of other assignments and in stolen moments of time. Working on it forced me out of my comfort zone in a number of ways and compelled me to rely extensively on the wisdom and kindness of others—on scholars who did the basic research that underlies my arguments, on scientists and technologists who explained their work to me, on the many people who described their experiences with reproductive and dating technologies, and on friends who patiently read drafts and listened to endless stories of life on the new frontier. I am deeply grateful to them all.

On the expert side, I was fortunate to be able to speak with Cynthia Breazeal, Kenneth Hayworth, Martine Rothblatt, Anders Sandberg, Takanori Shibata, and several employees of Aldebaran, the French robotics firm. Although they may not agree with everything in this book, they were gracious in offering their time, their knowledge, and their thoughts. If they are the shapers of the future, I feel confident that we are in good hands. For help with the past, I am similarly grateful to Caroline Elkins, Helen Fisher, Kimberley Patton, Sophus Reinert, and Richard Tedlow. I have undoubtedly made fewer mistakes in these pages than I would have without their help and patience.

I am also grateful to the wonderful group of research assistants who tackled various phases of this project: Aryanna Garber, who was with this book from beginning to end and embraced it wholeheartedly as her own; Lea Thomassen and Kalpana Mohanty, who dove into the enormity that is the Industrial Revolution; and Caeden Brynie and Julia Comeau, who stepped in to help with the avalanche of last-minute details. On the administrative front, I was extraordinarily lucky to have the support of Nadia Schreiber and Jamie Coffey. On the publishing side, I am deeply grateful to my agent, Suzanne Gluck, and to the incredible team at Farrar, Straus and Giroux. I owe a particular debt to Sarah Crichton, who first believed in this book, and to Alexander Star, editor extraordinaire, who took it on and shaped it powerfully. I owe a huge thanks as well to Nitin Nohria, the dean of the Harvard Business School, who welcomed me back so warmly in 2018 and made me feel at home.

As *Work Mate Marry Love* progressed, many friends and colleagues read it, critiqued it, and helped make it better. Noreena Hertz and Allison Stanger were vital sounding boards throughout the book's development, as they have been so frequently in the past. I am grateful for their counsel and for their friendship. Jennifer Finney Boylan was a brilliant and compassionate reader of my trans material, and Lura Chamberlain helped me rethink my exploration of masculinity. Along the way, Stephen Barden, Patrick Chung, Jane Isay, Jamie Metzl, Kalypso Nicolaidis, Bret Silver, Nate Snyder, and David Welsh offered wisdom, insight, and good humor; Kit Haggard and Ashiana Jivraj ushered me gently into the lives and mindsets of the millennial generation.

Finally, my greatest debts are those that lie closest to home. And although much of this book's argument centers on the evolution and possible disappearance of the nuclear family, it is precisely, if ironically, my own family that has been the foundation of my support. My parents, Judy and Marty Spar, and my brother, Marc Spar, cheered me on through the marathon of writing; my children, Daniel, Andrew, and Kristina Catomeris, patiently read drafts, offered opinions, and suffered good-naturedly through endless dinner conversations about love, death, and sex. In the course of this book's evolution, my sons also brought their new loves into our family, and it has been a mother's joy to welcome Lura Chamberlain and Carolyn Murphy into the clan. As always, though, my biggest debt is to my most loyal fan: Miltos Catomeris held my hand through this book as he has through so much else. If I am optimistic about the fate of love in a changing world, it is almost certainly because of him.

Index

Page numbers in *italics* refer to illustrations.

death (*cont.*)

immortality and end of, 239–64; leading causes of, *339n9*; life span and, 241, 245, 249–50, 259–61, *338n4*, *338n7*; memories and, 239–40, 241, 254–55, 263–64; mortality and human condition, 240–41; Ponce de León and, 246–47, *340n21*; rates, 171–72, *326n52*; religion and, 261; social structures and, 240–41, 261–62; soul and, 247, 251–59

debt, *338n6*

de Grey, Aubrey, 250

Descartes, René, 251–53, *252*

Dialectic of Sex, The (Firestone), 266–67, 273

digital immortality, 247, 254–59, 262–63, *342n47*

domestic revolution: housewife and, 56–63; Industrial Revolution and, 41–42, 55–56, *302n55*

dopamine, 147–49, 245, *319n58*

Doudna, Jennifer, 283

Douglas, William O., 107–108

dowry, 27

dryer, 76, *306n31*

dualism, *252*, *252*–53, 264

"Dutch protocol," 201–204, 206

E

economy: automation and, 156–59, 162–64, 225–26, 284; automobiles and, 72; debt and, *338n6*; education and, 164; in future, 268–73; gig, 270–72, *345n5*; household appliances and, 79–80; during Industrial Revolution, 59–60, 62–63, *303n57*; labor and, 161–63, 175–77, 225–26, *338n6*; masculinity and, 156–57, 175–77; robots and, 225–26; sex and, 128, 144

education: economy and, 164; feminism and, 167; gender studies and, 167; labor and, 164; marriage and, *320n67*; women and, 167, *324n36*

Edwards, Robert, 98–99, 103

eggs, *see* reproduction

Elbe, Lili, 198, 204

electricity, 214, *333n8*

Eliot, T. S., 242

Empire of Cotton (Beckert), 45

endocrinology, 191–96; *see also* hormones

End of Men, The (Rosin), 170–71

Engels, Friedrich, 10, 20–21, 33, 40, 55, 266

Enigma code, 216–17, *217*

eonism, 191

Ephron, Nora, 275

Epic of Gilgamesh, 246

ethnicity, 130, 139–41, *318n34*, *318n38*

extramarital affairs, 29, 34–36, 149–50

F

factories: automation and, 53, 57, 162–64, *321n4*; capitalism and, 52; cotton, 51–52, 57; factory girls, 56–57; labor, 51–55; rise of, 51–53; robots and, 156–59, *157*

hackers on, 247–50, 264; death and, 239–64; digital, 247, 254–59, 262–63, 342n47; dreams and myths of, 240–41; fountain of youth legend and, 246–47, 340n21; future and, 261–62

immortal jellyfish, 248

impotent men, 97–98

incels, *see* involuntary celibates

income gap, 226, 323n23

Industrial Revolution: agriculture and, 42–45, 57, 63–64; automation and, 53, 57, 224–25; capitalism and, 40, 41; change and, 63–66, 88; domestic revolution and, 41–42, 55–56, 302n55; economy during, 59–60, 62–63, 303n57; factories and, 51–53; home and, 39–41; the housewife and, 41–42, 56–63; labor division during, 41, 53–56, 64–65, 158–59, 300n17, 302n43; machines and, 9, 39–66, 224–25; marriage and, 41–42; nuclear family and, 58–59, 302n55, 303n60; origins of, 42–45; production and, 39–41; robots in, 214–15; social structures and, 55–56, 61–63, 88, 130, 158–59; steam during, 39–42, 45–50, 52–53, 64, 300n22; Watt and, 40, 45–50, 50, 52–53; women in, 9

inequality: income gap and, 226, 323n23; technology and, 283–87

infertility, 97–98, 309n4

internet, *see* online dating and sex

internet addiction, 319n56

intra-cytoplasmic sperm injection (ICSI), 95

in vitro fertilization (IVF), 10, 11, 98–104, 116

in vitro gametogenesis (IVG), 116–22, 314n65

involuntary celibates (incels), 144

iPhone, 133–34; *see also* online dating and sex

IVF, *see* in vitro fertilization

IVG, *see* in vitro gametogenesis

J

Japan, 143, 147, 151, 315n11; life span in, 241; robots and aging in, 229–31, 336n34

Jensen, Peter, 27

Jeopardy!, 221

Jetsons, The (cartoon), 122

jobs, *see* labor

Johnson v. Calvert (1990), 111

Jorgensen, Christine, 330n38

K

Kahn, Irving, 245

Kasparov, Garry, 221

Kass, Leon, 99

Katz, Lawrence, 87

Kay, John, 48–49

Keynes, John Maynard, 155–56, 177, 270, 273

Kimmel, Michael, 171–72

King, Billie Jean, 115

Klinefelter syndrome, 189

Knowlton, Dr. Charles, 84

Krause, Harry, 108, 111

Neolithic Revolution, 17
Netherlands, 182–83
Newcomen engine, 46–48, 47
Nietzsche, Friedrich, 239
non-cisgender individuals, 206
North, Douglass, 24
nuclear family, 6, 8, 58–59, 302n55, 303n60

O
Obergefell v. Hodges (2015), 115
online dating and sex: beginnings of, 131–34; brain and, 147–49; choice overabundance and, 141–42, 147–50; ethnicity and, 139–40; gender and, 150–51; with Grindr, 126, 133–34; Happy Families Planning Service, 131–32; history of, 129–35; homophily in, 337n27; hookup culture and, 317n28; involuntary celibates and, 144; iPhone and, 133–34; love and, 142–43, 152–53; marriage and, 131–33, 149–51; with Match.com, 132–33, 136–37, 140, 316n17; monogamy and, 148, 149; with OkCupid, 320n69; sex frequency and, 127–29, 143, 145, 149–51; social structures and, 126–27, 139–40; with Tinder, 125–29, 134–40, 145, 150–51, 275, 316n22
oxytocin, 148

P
Pancoast, Dr. William, 97–98, 103
Paradise Lost (Milton), 209, 332n1

parenting: laws and, 105–10, 312n50, 315n73; "multiplex," 120–22; poly, 116–22, 315n73; same-sex, 104–106, 112–16, 120–21, 312n46, 312n50
Parkinson's disease, 245, 319n58
paternity, 26, 36, 94, 106–10
patriarchy, 26–28, 34–36
pension plans, 345n7
PGD, *see* pre-implantation genetic diagnosis
Pincus, Gregory Goodwin, 86
Pindar, 212
pituitary gland, 193–94
Place, Frances, 83–84
plantations, 42–43
Plato, 252
plow, 8, 63; Ancient Egyptian "scratch plow," 23; creation of, 22; in late Predynastic to early Dynastic Period in Egypt, 21; remaking of ancient world and, 21–31; social structures and, 22–24
polygamy, 8, 27–28, 293n8
poly parenting, 116–22, 315n73
Ponce de León, Juan, 246–47, 340n21
Pope, Barbara Corrado, 60
population expansion, 31, 33, 43, 297n53
pornography, 128
potato farming, 42–43
pre-implantation genetic diagnosis (PGD), 101–103, 118, 119
property ownership, 298n70; agriculture and, 24, 26–28, 35–36, 298n71; patriarchy and, 27–28, 35–36
pruning, synaptic, 339n14
purpose, and labor, 226–27

trans revolution: bathrooms and, 203; definitions, words, and "trans," 185–87, 207; "Dutch protocol" and, 201–204, 206; feminism and, 331n46; gender and, 181–85, 276–77; gender dysphoria and, 182–83; Hirschfeld and, 196–98, 203–204; history of the third sex and, 182–83, 187–90; hormones and, 183, 190–96; Money and, 200–201, 331n41; science, gender, sex, and, 183, 190–204; Steinach and, 194–97, 329n25, 330n30; stories, 204–207; surgery and, 196–99; transitions and, 181–85

triangle trade, 44

Trump, Donald, 284, 326n53

Turing, Alan, 216–18, 334n16

Turing test, 218–19, 334n16

Turkle, Sherry, 234–35

Turriano, Juanelo, 213

U

Uber, 270, 345n5

Ulrichs, Karl Heinrich, 191

unemployment, 165–66, 322n7

United Therapeutics Corporation (UTC), 253

universal basic income, 226, 335n31

unmarried mothers, 6

V

vasopressin, 320n59

Venuses, 29, 30

Vinge, Vernor, 222, 239

violence, 144, 339n9

virginity, 27, 34, 35, 84–85, 127–28

virtual sex, 233–34

W

wages, 60, 226, 323n23, 335n31

washing machine, 76, 78, 165

water, 32

Watt, James, 40, 45–50, 50, 52–53, 64

weapons, AI, 222

Weigel, Moira, 148–49

When Harry Met Sally, 275

"white goods," 79–80

Whitehead, Mary Beth, 110–11

Winston, Robert, 101–102

Wizard of Oz, The, 215, 229

Wolf, Naomi, 78

women: agriculture and, 8, 25–31, 35–36; ancient ancestor, 29, 296n37; in antiquity, 33; automobiles and, 70–73, 306n20; death rates for, 326n52; education and, 167, 324n36; factory girls, 56–57; feminine ideal, 60–61; housework by, 62; identity and, 169–70; in Industrial Revolution, 9; labor and, 167–68; religion and, 17, 29; reproduction and, 94–95, 97–98, 106–10, 117–18, 313n61; Venuses, 29, 30; worship of, 29; *see also* birth control; feminism; housewife; labor division; reproduction

Woolf, Virginia, 59–60

work, *see* labor

World's Fair, 1964, 4

World War II, 216–18

A NOTE ABOUT THE AUTHOR

Debora L. Spar is the MBA Class of 1952 Professor of Business Administration at Harvard Business School and the former president of Barnard College. Her previous books include *Wonder Women: Sex, Power, and the Quest for Perfection* and *Ruling the Waves: Cycles of Discovery, Chaos, and Wealth from the Compass to the Internet*.